Exploration in Science

in your classroom

Edited with an Introduction by
Julie Smart

Series editors:
Tony Martin, Mick Waters, Wendy Bloom

© Mary Glasgow Publications Limited 1991

First published in 1991 by
Mary Glasgow Publications Limited
131–133 Holland Park Avenue
London W11 4UT

Illustrations by Nancy Martin
Designed by Dave Sumner
Typeset by Euroset, 35–37 Essex Road, Basingstoke, Hants RG21 1TB
Printed by The KPC Group, Ashford, Kent

British Library Cataloguing in Publication Data
  Managing exploration in science
  1. Schools. Curriculum subjects: Science. Teaching
  I. Smart, Julie
  507.1

Contents

Exploration in Science
in your classroom

About the series editors

Tony Martin is a Lecturer in Language and Reading at Charlotte Mason College of Education, Cumbria. A former deputy head, he now runs in-service courses for primary teachers across the country, and speaks regularly to teacher and parent groups. He has written a book called *The Strugglers* about his work with children who have difficulty learning to read.

Mick Waters is a Lecturer in Primary Education at Charlotte Mason College of Education, Cumbria, and a former headteacher. He is responsible for running in-service courses for primary teachers and heads across the country. Many of these courses address the issues which are the focus of this series.

Wendy Bloom is Co-ordinator for Language in the Primary School at the College of St Mark and St John, Plymouth, and has wide experience of teaching in schools and in-service courses. She has published a book on parent partnership in reading, worked with the BBC and Open University on this topic, and contributed to books on other topics.

They are also co-authors of the following titles:

Managing to read: a whole school approach to reading
(Mary Glasgow Publications, 1988)

Managing writing: practical issues in the primary classroom
(Mary Glasgow Publications, 1989)

About the contributors

Julie Smart is a freelance educational journalist with a special interest in primary science. From 1988 to 1989 she was editor of *Questions*, the science magazine for primary teachers, and she has also worked for *Junior Education* and *Child Education*.

Peter Griffin is headteacher of Doe Bank JMI School in Walsall.

Neil Burton, a former primary science advisory teacher, is now science co-ordinator at Pegasus JMI School in Birmingham.

Di Stead is science co-ordinator at Limpsfield Community Middle School in Sheffield.

Martin Skelton is a former primary headteacher and now an educational consultant in Ipswich.

Corinne Murray is science co-ordinator at Sandylands County Primary School in Morecambe, Lancashire.

Mike Sullivan is headteacher of Busill Jones Primary School in Walsall.

Introduction

The development of science in the primary school curriculum promises to be one of the most positive innovations of the 1990s. As schools throughout the country embark on new initiatives and teachers tackle the subject with renewed enthusiasm, more children than ever are enjoying the excitement and challenges that science has to offer.

For this book, we have invited a number of primary practitioners, like yourself, to share their ideas and experiences. If you are a non-specialist, we hope it offers some help and encouragement in areas you may feel uncertain about. If you are already an enthusiast, you may prefer to use the contributions as starting points, as suggestions to try or adapt according to your needs, or perhaps as a basis for discussion with colleagues.

The chapters set out to address five major concerns about science which are uppermost in teachers' minds.

- How do I manage practical science in the classroom so that every child is offered the best possible experience?
- Where do I start if we don't even have the resources?
- How can I organise topic work?
- What about assessment?
- What could we be doing as a school?

No one can give you the answers. Every school has its own approach, all teachers have their own style and way of doing things, and so this book cannot provide a single model that is appropriate or relevant for everyone. What it attempts to do is highlight the key issues involved in managing exploration activities in science, and guide you through the challenges which may lie ahead.

Managing science in the classroom

So, how do you manage practical science in the classroom so that every child is offered the best possible experience?

There is a certain mystique about science. It is often seen as complex and technical when in fact it is simply a way of ordering and explaining everyday phenomena. Are your doubts really about science, or are they more about the problems of practical work?

Managing practical work brings many pressures and Chapter 1 looks at this area in detail. Everyone has their own teaching style and you cannot expect to change overnight. What you can do is identify that style, use what you are comfortable with and plan the children's experiences accordingly. In this way, you only have the science to worry about and not the organisation as well!

Have faith in your own skills and experience as a teacher, and your knowledge and understanding of young children. Good practice hasn't changed. A good teacher is a good teacher, whether you know about science or not.

Once you overcome your fears and inhibitions and realise that you are not being asked to abandon all that you've learned and start from scratch, you can begin to be positive. The first and biggest hurdle to overcome is learning to voice your concerns and share your problems. You will soon find out that you are not alone!

Where do you start?

No school ever seems to have enough resources, and lack of equipment can be a major constraint on practical science. How many schools have a binocular microscope in every classroom, or enough hand lenses to go round? Even if they have them, are they accessible, serviceable and properly used?

Unfortunately, it is usually a case of making the most of what you do have – something with which primary school teachers are all too familiar! Chapter 2 offers some ideas on how to go about acquiring and organising resources.

Do not be misled by the secondary school science image of test tubes, bunsen burners and sophisticated equipment. Apart from a few special items such as bulbs and batteries, you can probably present most National Curriculum science with everyday household items. You'd be surprised what you can do with ordinary school stock, and a few cotton reels and yoghurt pots!

However, you do need to cultivate the right atmosphere for learning, whatever the curriculum area. Do children know where to go for different things? Are they taught to return them? Can they be encouraged to find things they need without wandering around the classroom or disrupting classmates?

Don't try to hand over complete responsibility without careful preparation. Children need time to adjust to new approaches and you must be prepared to go slowly at first if they are not used to working in a certain way. They will improve once it is no longer a novelty.

The same applies when introducing a class to something new, such as group work or problem solving. Decide where your priorities lie – is a calm and peaceful atmosphere the only learning environment, or do you really want the children to learn to discuss, debate and talk through their ideas together sometimes?

The secret is to think through beforehand what could possibly happen and be prepared. Make sure as far as you can that there are enough materials for everything you want to happen to happen, and share your thoughts with colleagues who may have different experiences to contribute. You cannot be expected to anticipate everything!

Organising topic work

Topic work is widely accepted as good primary practice. At its best, it can provide a meaningful context for science and offer children exciting opportunites to learn about the world in a relevant way.

On the other hand, cross-curricular topics which bring in too many aspects of different subjects can lack cohesion and provide children with a superficial experience of science (HMSO, 1979).

A popular approach is to choose curriculum-based topics to ensure that key areas are being covered. For science, that also means choosing a theme for the processes and skills it offers as much as for its content.

Continuity and progression must also be taken into consideration when choosing topics and so it makes sense to plan co-operatively as a staff. Some schools are making topic work a whole priority, allocating one teacher responsibility and time for cross-curricular planning, perhaps with reference to a specific area such as science and technology. Many benefit from planning and evaluating in year groups.

More of the options available are discussed further in Chapter 3, which focuses on managing science in the whole curriculum.

Assessment

Whatever your style or approach to teaching science, matching the level of activities to the ability of *individual* children is crucial.

Chapter 4 looks at the whole issue of assessment but, remember, you know your children. You are probably used to keeping a lot of information about them in your head. Although recording increases the demands placed upon you, having access to specific and accurate information can only add to your confidence and professionalism.

Managing science in the school

Most schools now have a person with responsibility for science, or a co-ordinator who you can approach for advice or who may even be able to come into your class to lend a hand. Unfortunately, few of them have the time or opportunity to co-ordinate. Time simply isn't built into the school day for discussing problems in any depth.

Chapter 5 examines the role and responsibilities of the headteacher and identifies some of the key areas which need to be covered by a science post holder.

If science isn't a priority in your school and you think it needs to be, then there is probably very little you as an individual can do to change things. Staff can only move forward with the backing of a supportive headteacher who can motivate people and manage time in a way that is beneficial to the whole school. However, you can show through your classroom practice the value of a good science education.

Teaching science to young children is tremendously rewarding. It gives you fresh insights into your children, their qualities and their characters. It gives them a chance to investigate the world around them and experience the excitement of discovery. So, enjoy science, encourage your children to enjoy science and seize this wonderful opportunity to learn and explore *together*.

① Managing Science in the Classroom

Science teaching has a tradition of practical activity, specialised equipment, academic rigour and specialist teachers. All of these can conspire to make infant and junior teachers feel less confident about organising science lessons.

There is no one style of classroom management that will suit:

- all science teaching situations;
- all children in all classes;
- all teachers in all schools;
- at all times.

Peter Griffin describes how a good teacher can use an amalgam or variety of management styles to suit differing circumstances.

Certain ways of managing the classroom are more commonly used than others depending on the objectives behind teaching science. For these we can look to the National Curriculum, which firmly places an emphasis upon process-oriented, activity-based science. By calling Attainment Target 1 'Exploration of Science' (which is a profile component in its own right and carries the same weighting as ATs 2–17), the suggestion is that children should generally 'explore' science through a practical not a prescriptive style of teaching. The requirement to maintain a differentiated curriculum – which matches the work to the child – suggests the need for some form of individual or group work. The single most appropriate style of classroom management would seem, therefore, to be one that enables children to learn at their own individual level, through open-ended practical work. However, this approach may not always be the most appropriate nor even possible in the confines of the classroom.

When deciding which organisation is appropriate, you have to take account of:
- the differing needs and abilities of children;
- the nature of what is being taught;
- the efficient use of teacher time;
- the style of the teacher;
- the nature of the school;
- the resources available;
- the time available.

There are three basic types of organisation in which the children are taught:
a whole class; a small group; individuals.

Whole-class lessons

Generally, these are seen as relatively easy to organise and they seem to make fewer demands upon teacher time and preparation. However, there are two extremes. On the one hand, there are the prescriptive 'chalk and talk', demonstration experiments and 'experiments by numbers'. On the other, there is the whole-class lesson which tends to be more flexible and open-ended but often develops into a group or individual activity.

Chalk and talk, **demonstration** and **experiments by numbers** seem useful at first sight because the whole class is focusing on the same thing at the same time. However, even in the 'experiment by numbers' lesson, in which pupils are told exactly how to do an experiment to prove a certain point, the children are not actively involved. They take little part in the decision making and the learning challenge tends to be passive. Such lessons rarely take account of the different needs and abilities of the children. However, the whole-class lesson has some value if it is properly prepared as a stimulus or starter to wider-ranging activities.

First, it is easier to ensure progression and continuity. Second, the demonstration lesson, if selected with care and presented well, can be entertaining, provoke thought and investigation, and form a stimulus for other, more active lessons.

An opportunity for stimulating the whole class in this way arose when some children I was teaching were germinating seeds. I wanted to help the children understand controls (AT 1) in an experiment and fair testing, so I took a germinating bean and sang to it three times a day. The bean grew, apparently proving that singing makes beans grow. But the children knew this was outrageous, and through their own experiments began to grasp the concept of fair testing.

Television programmes, visiting speakers, film slides and artefacts such as working models, posters or samples of materials are all useful for stimulating interest and response.

A **whole-class lesson** is useful for setting the scene at the start of a science project. A class discussion may take the form of a think tank in which random ideas are initially put forward. This 'starter' lesson can help you and the children to organise the next few lessons. The work to be done can be negotiated and resources collected together. You can also anticipate and highlight some common problems that the children may come across. These may include establishing safety rules, explaining new terms or making clear each group's aim. The whole-class lesson is also useful for drawing a project to a close, or in an interim stage as a plenary session.

Often what starts out as a whole-class lesson can become more individualised. If the class is directed in such a way that children can actively make differing responses and go off at viable tangents, a lot of learning experiences can occur. For instance, in the programme of study for AT 11, top infant or lower junior children make a simple circuit. You can either teach this as a skill which requires little understanding (experiment by numbers), do it through some form of group work, or present it initially as a challenge to the whole class. The advantage of the challenge is that it introduces a topic quickly and allows you to see what previous knowledge and expertise children are bringing with them. However, it does require a lot of resources and, if not carefully planned, can degenerate into chaos.

The challenge could be to give children a battery with brass strip terminals (PP3) and a bulb and see if they can light the bulb. Eventually, by trial and error, all the children will manage to light the bulb. Then gradually, with a limited amount of discussion, introduce them to a bulb holder, then a wire and then two wires. The challenge is always to find as many different ways to light the bulb as possible. Afterwards this can be extended individually, in pairs or groups, into other circuit work.

The whole-class lesson, therefore, is of limited value for the exploration of science, unless used as an initial stimulus or organisational foray. It can have a further value when some kind of rote or elementary learning is required. This is especially important if you need to make clear certain safety or organisational rules.

Whatever the style of general organisation, the whole class does have to adopt a routine for getting out equipment, putting it away and storing unfinished projects. In many schools, science equipment is stored in a central place. School policy will determine if and what children can fetch and carry. A well-organised teacher will anticipate most of the equipment needed and make sure that it is available in the classroom. The best motto is: 'There's a place for everything and everything in its place.' You can help by labelling the places for equipment. If boxes are not used, paper mats can define the area for each item. Some teachers prefer to use monitors to give out and put away equipment, while others prefer the children themselves to assume responsibility. The former is often easier to organise, although the latter should be part and parcel of a child's self-education. If the children are trained to select, seek and clear away their own materials, not only is it good for them but it also frees you for more important work.

It is easy to dismiss the whole-class lesson and activity as outmoded and old-fashioned. However, it is a useful way of organising a class for some activities. To totally dismiss it would denigrate the richness of experience and language children get by clustering around the teacher armed with a big book, poster or an object brought in by a curious child.

Group and individual work

There are many ways of organising group work. One method (covered on page 15) involves the whole class doing a similar activity, with the teacher maintaining a balance between total direction and open-endedness. The children work individually or in groups. Such a pattern is relatively easy to organise, the groups generally doing similar activities and moving towards similar goals. It is easier to keep track of how groups are progressing. You can identify potential difficulties and direct the class as a whole body. In the earlier example about electricity, for instance, you might show the whole class how to bare a wire and attach it to the bulb holder. Otherwise, if groups of children are doing this at different times, it means that you will have to repeat yourself up to nine times. If every group is doing a similar activity, and the classroom atmosphere is positive enough, groups can bounce ideas off each other and help one another when problems arise.

The problems associated with such an approach could totally devalue it without proper planning. If there are not enough resources for all groups to be working effectively all the time, then they will stop and lose interest. If the approach is not sufficiently open-ended, so that children can work at their own level, then the activity may not be demanding enough of some children, or too demanding of others. There is also the danger that one group will inadvertently copy another and you will assume that the children understand what they have done. If the classroom atmosphere is over competitive, some groups will inevitably be seen as more successful than others. On the other hand, although one group may have successfully completed an experiment or investigation, another may have encountered innumerable difficulties and in so doing learnt more about the process of science.

A further problem associated with group work is mixed-ability teaching. In an attempt to counter the poor self-esteem of a low attainer, teachers often put children in mixed-ability groups for science. However, in this situation less able children will often not participate fully in group work. They may end up doing some menial task or may even mimic the rest of the group to appear successful. In science there are opportunities for mixed-ability grouping providing you know the strengths of individual children. Mixing children with different talents in order to form a cohesive group, especially in problem-solving activities, can achieve a lot.

Placing a slow-learning child in a group where he or she is competing directly with high academic fliers does little to improve the self-image or the class-image of that child. Perhaps what is more likely to improve self-image is to put the child in a situation in which he or she can contribute and gain success. This may at times mean streaming groups and differentiating the tasks. Providing this is handled with sensitivity, and is not the only way of grouping, it should be no more damaging socially than any other type of organisation.

Another alternative which raises self-esteem is to involve children in helping to run groups for younger pupils. In reality, putting children, regardless of ability, in a group where they are going to be successful and where their contribution is as valuable as any other group member, is more likely to have a positive social effect. The less able group may be given more teacher support, with specially designed work cards and record sheets to help them perform the same tasks as other children.

Alternatively, they may be given an open-ended task which can be performed at different levels. For instance, if lower juniors were looking at activities in the programmes of study for the 'Processes of life' (AT 3) and the 'Variety of life' (AT 2) they might design some experiments to do with the conditions needed for germinating seeds. The more able children, higher up Key Stage 2, might design sets of experiments to explore a variety of conditions, using variables and large samples for testing. To see the effects of temperature they might test large numbers of seeds at different temperatures and make sure that all other possible variables such as light, moisture and growing medium are the same in each test. They might plot the ratio or percentage of seeds germinating and draw conclusions from this. Less able children may do a similar test, testing fewer temperature regimes, taking little account of other possible variables and not subjecting the results to any statistical analysis. Their observations may be done by simple visual comparison. One group might produce reams of writing, graphs and tables, while the other may record the experiment with pictures or photographs, with pressings of the germinated seedlings as part of the conclusion. If both are seen as equally viable by the teacher and given equal status, then the ideals of mixed-ability teaching within the classroom are often more closely achieved.

Having the whole class, even in groups, doing science at the same time is not always best. Some teachers, especially of older juniors, may prefer to have a timetabled science lesson. Others may prefer to have some form of integrated day. There are various ways of organising this. Some groups of children may be doing practical science work, while others are doing quieter, more independent tasks. During a week each child or group of children rotates through a variety of tasks, one of which is science. This is a pattern of organisation which particularly suits some teachers with some classes.

Although more popular in infant classes, it is equally successful in junior classes. Such a system is especially appropriate for team teaching, where there is more flexibility for different types of groupings. For example, one teacher can look after one large group of children while another can do some in-depth science work with a smaller group. An integrated approach does free a teacher to give a group of children doing science some extra attention and offers the opportunity to do thematic work.

For some science activities, such a pattern of organisation may be absolutely necessary. If the children are doing a project which requires extra supervision, you will have to find some way of giving a group additional attention, for example, when children are using craft tools such as a glue gun. You may give the rest of the class some useful but quiet activity or get extra help from an ancillary, 'floating' teacher, a parent or even the headteacher!

You may also need to find individual time for a group or child who is having some specific problem, or for assessment purposes. It may be necessary to adopt this pattern of organisation because there are only enough resources for one or two groups to do a task at one time.

This problem, however, does not arise in classes that are organised on an individual assignment basis. To be successful, such an approach requires a high level of organisation, with you keeping track of a lot of groups doing a lot of different things. It also requires a high degree of independence on the children's part. The danger is that you become merely an administrator, and while the children might be self-motivated to do a lot of work, because of the lack of teacher intervention and direction there is little progression or quality. Such a system requires a teacher capable of being well organised, with the ability to move purposefully from one thing to another, and by constant monitoring and assessment ensure that individual children follow a progressive programme of development. Such a

system has the potential to be highly successful but is also capable of breaking down into chaotic mish-mash without proper organisation.

Circus of experiments

A compromise between this and the whole-class thematic approach mentioned earlier is the circus of experiments. A set of assignments, based on a theme such as 'Light' or 'Ourselves', should be placed around the classroom. Always include one or two extra assignments, so that there is a spare activity for a group that works fast.

There are many permutations. For instance, for a class of thirty-two divided into eight groups of four, there may be nine different assignments or five assignments with enough materials for each to be shared between two groups. Collect together equipment, materials, fact books and work cards, possibly in a box for each assignment. Then give these out at the beginning of each lesson.

Groups rotate from one assignment to another, reporting back to the others when they have completed a task. In this way, one group can forewarn another of possible problems. In the busy classroom, children who have completed an assignment may occasionally act as trouble shooters for other groups. In such a system children need to be prevented from seeing the assignments as ends in themselves. They should still be positively encouraged to question and allowed to go off at a viable tangent.

As with any of the organisations mentioned the crucial elements are that the system has to promote continuity and progression through science work, while being adaptable enough to take account of children working at different levels of understanding.

Some teachers try to achieve this balance, by any of the systems mentioned, around a science table. This can be a display of artefacts, equipment and assignment cards. Throughout a week, groups of children take turns in working at the table. Only by teacher intervention or by altering assignment cards can the needs of each group be met. Alternatively, the science table can be the focus of attention for a class theme during a whole-class lesson or on an integrated day basis. In this way, its main function is not to act as a work station but as a stimulus for a project, storage area for equipment and display space for children's work.

Whatever classroom organisation is adopted, much can depend upon how well organised the classroom layout is. A classroom is rarely ideal for all the activities that go on there. However, teachers can avoid a lot of disturbance if the room is adapted as far as possible to suit the activities going on. For group work tables need to be put together to form large, flat surfaces. Some teachers, especially in open-plan schools, find that they can make bays for groups to work in, and store equipment and unfinished work. If the classroom does not have enough storage space, it is worthwhile investing in large plastic stacking crates. They are useful for storing a group's work, or the assignments for the circus organisation. However the classroom is organised or laid out, it is important that children have a set routine for accessing equipment and resources, and for clearing away.

Recording and monitoring

It is also important that you maintain a good system of recording and monitoring children's work. This is important not only for final assessments, but also helps with organising groups and matching work to the ability of the children. It is sometimes possible to combine short term planning and recording. Each group could have a sheet, listing their assignments down the side and the names of the group members across the top. Where the name of a child intersects with the assignment, you can record observations about the child. In this way, you can keep a general record on each group and also specific details about an individual in the group. Such a sheet helps with the day-to-day running and planning of group activities.

There is no one form of classroom organisation that covers all eventualities. Some methods are more appropriate to active science work than others, namely those that allow children to work in groups, or as individuals. The organisation needs to accommodate children of mixed ability, and yet has to be sufficiently manageable so that you can monitor individual children's progress. The best approach is to identify the teaching styles with which you are most comfortable and plan the children's activities accordingly. Then decide upon the most appropriate pattern of organisation and monitor the children's progress.

An example – circuits

A flexible pattern of organisation over a number of lessons might run in the following way.

In a topic on electricity, you would need at some point to do a whole-class 'chalk and talk' lesson on electrical safety. You might make this the time to set the scene, perhaps by presenting the children with an open-ended challenge (see page 15), brainstorming, or using stimuli such as a science table of artefacts, posters, a TV broadcast or film slides.

From these, the children may be given individual assignments or work in groups. If the groups are given the same sequence of activities – say, making and explaining increasingly complex circuits – some children will need additional challenges, others will need extra help.

Those who need support could be given a worksheet showing different ways of connecting wires, batteries and bulbs. You might ask them to guess whether each circuit will light the bulb before they try it. Other children might go beyond this, inventing their own circuits.

Children who manage quickly to build the simple circuit, and seem to understand what they are doing, could broaden their experience with a circus of experiments. For example, they could apply their skills to lighting up a model house, or a sign for display. Challenge them to design and make their own bulb holders, switches and battery holders. By using reference material, they might extend their new-found understanding to explore circuits in the home.

Alternatively, rather than give the more 'electrically' able children a wide base of experience, you might give them more complex work, asking them to build circuits with two or three lights, incorporating switches in different positions.

Whatever the pattern of organisation, there should be some opportunity for groups to demonstrate what they have found out to the rest of the class. You will also need to bring the whole class together to bring the activities to a conclusion, to assess what the children have learned and to give the topic an overall perspective.

No one can dictate what is best for you or your class. You know your children, their strengths and weaknesses, how they interact, where they lack confidence, what fires their imaginations. By keeping an open mind about what approach to follow and responding in the way you feel is right, you can help your children to achieve their potential.

Ideally, each child should have an individually programmed science education. Being responsible for a class or group of children makes this virtually impossible but, by adopting a variety of management strategies, it is possible to get closer to this ideal.

❷ Managing Your Resources

Unless your school acquires and manages its science resources effectively and efficiently, they may be a waste of school funds. It is all too easy to misuse, lose, ignore or simply squander the school's resources. Neil Burton's chapter will help you to identify and maintain your science resources as well as offering a few ideas on how you may wish to organise them to suit the particular needs of your school.

Making an inventory

Before attempting to organise, store or even acquire science resources you should always make an inventory of what you already possess. In making this inventory you should also make note of the 'type' of resource and how much it is used. Is it underused because it is not very useful, because no one appreciates or understands how to use it, or because nobody knows it's there?

I find it useful to divide resources into eight categories:

- capital items of equipment;
- consumables;
- 'junk';
- materials for investigation;
- printed resources;
- environmental resources;
- audio-visual resources;
- human resources.

Capital items

These are often the most easily identified as being 'science equipment'. Included in this category should be:

- optics (hand lenses, mirrors, prisms, torches);
- magnets (alnico and bar types, compasses);
- electrics (batteries, bulbs, motors), though these should be considered as semi-expendable;
- containers (plastic aquaria, pooters, Pyrex saucepan, bucket);
- construction kits (Lego, Meccano, LegoTechnic, First Gear);
- measuring apparatus (for time, distance, weight, force, temperature, weather, capacity/volume);
- any other major piece of science equipment you may possess (model skeleton, ant village).

Consumables

The number of different items in this category are almost too numerous to mention. The following is a basic list:

- kitchen chemicals (dried food stuffs, dyes, sugar, salt, washing-up liquid, safe (non-bleach, non-biological) washing detergents);
- plant growth (seeds, compost, rooting compound);
- kitchen materials (tin foil, cling-film, kitchen roll);
- bags (paper, plastic, polythene, bin bags);
- classroom materials (art straws, drinking straws, adhesives, art materials, modelling materials, paper clips, string);
- science materials (iron filings, bell wire, filter paper).

'Junk'

This category is as useful as you make it. Depending on how good you are at finding alternative uses for objects such as plastic lemonade bottles, your store of junk could be anywhere between your most valuable resource and a staging post between parents' dustbins and the local dump. In other words, only collect things for which you can see several uses, or at least one very good use! Junk that I've found useful over the years includes:

- jam jars;
- clear plastic bottles;
- yoghurt pots;
- newspaper;
- 35 mm film containers (the clear ones make good specimen pots and the others good battery holders);
- white margarine or ice-cream tubs (for viewing minibeasts).

Once you start to look for the alternative uses of common disposable objects you will surprise yourself with how many you find.

Materials for investigation

In many ways this category is similar to the previous one, except you are collecting materials based on what they are made from rather than for their potential use. The sorts of things you should be looking for here are plastics, fabrics, strings, threads, papers, packaging (polystyrene, corrugated paper), floor coverings (tiles, carpet), elastic bands, and also natural materials such as soils, stones, and tree barks. Some other materials that can be purchased as sets are metals (for magnetism and electricity), woods, rocks and minerals.

Printed resources

For a list of useful resources see pages 68–73. Books, magazines and posters are invaluable as means of providing a stimulus for both staff and pupils in addition to their usual use as research material. The science section in the school library should not be ignored, especially by yourself, as these books are often the pupils' first point of reference and their questions will often be based on them. Story books are also an excellent means of stimulating investigations or reinforcing scientific principles. It is worth noting any subscriptions the school has to scientific magazines or organisations as they provide a wealth of ideas and information.

Probably your most prized printed resource will be your published scheme but please remember, whatever the publishers say, there will never be a perfect scheme. The nearest you will ever get to an ideal scheme for your school is if you write one yourselves. Knowing the school, the children and yourselves, you are the only ones in a position to do it. Pick the scheme with the structure you like (if any) and apply it to *your* topics with the activities *you* want. Don't let yourselves to be tied to one scheme; always allow a degree of freedom.

Environmental resources

One of the most important resources is the local environment, and by this I mean classrooms. They have windows through which pupils can observe nature in action (seasons, plants, animals, the sun, the weather), floor coverings, display boards, maybe even heating and water. Outside the confines of the classroom you have the school buildings and grounds. Even if the building is surrounded by tarmac and a high brick wall, you can make use of that (building materials), or take the opportunity to make the school environment more stimulating. Note where wild plants and minibeasts flourish within the grounds.

Beyond the school gates, where is the nearest park or natural wildlife oasis? From what materials are the local structures built? Are there signs of pollution? Even further afield, are there any places of interest that would make useful field trips, such as zoos, factories, nature centres, farms?

Audio-visual resources

Your school's audio-visual resources might well vary from a few free posters to a video camera. Others you may find include still cameras, cassette recorders and video (both with pre-recorded tapes), slides or film strips, and computer software (scientific simulations as well as data and word processor packages).

There is also a wealth of stimulating material being beamed into your school from both radio and TV. These programmes really need to be recorded so that you can use them when you want them, rather than just when they are transmitted. This will also allow you to interrupt the programme to question, discuss or review a portion of it, thus making it more valuable for the children in terms of developing their science skills.

Human resources

The most vital component of any organisation is people, especially in schools. Primarily the expertise of the teaching staff should be recognised – most practical hobbies (or chores) have a scientific basis. Anything from washing-up to bee keeping could be of value in a classroom. Don't forget the skills of non-teaching staff, from the listening skills of the nursery nurse, to safety and nutrition information from the kitchen staff. You may well have parents who would like to take a class on a field trip, or listen to the children's ideas in the classroom. Parents, governors or indeed any other members of the local community may be able to furnish the school with specialised knowledge, equipment or places to visit.

Finally, for further help with your school's science you should not forget schools advisers and advisory teachers. They will be able to offer help with many aspects of the science curriculum, especially in the areas of resource acquisition and organisation.

Evaluating your resources

Once you know what you've got, you have to find out whether those are the things you need. If you already have a science scheme, be it your own or a commercial one, then it is a matter of matching your resources to the ones required by the scheme and filling in the gaps where necessary. Now isn't that easy! All you have to do now is find the money to buy them, find out how they work, only to discover that you could have bought them cheaper elsewhere and that nobody really wants to use them anyway because they are not exactly the same as the ones in the book!

Whenever possible go for advice to someone who has experience of buying for the scheme you wish to resource. Even if they can't wave a magic wand and solve your problems by doing it all for you, they may help you to avoid some of the pitfalls. The best advice I can give is, if you are unsure and it's expensive, borrow it from your local science centre first to see if it will be used at all. Don't try to get everything at once, just start with what you've got and buy if the demand is there, or if you're sure that you can create the demand.

Organising your resources

So you have identified the school's science resources and you have what you believe to be at least a basic kit to get you started. Now you need to organise them to ensure that they're used. A good science store is an empty one! Making sure that the equipment is being used and used appropriately comes later. Storage is a big problem as there are so many different solutions. Your 'answer' will depend on your school.

The first factor and, in most cases, the most important one is the science curriculum itself. For example, does your school organise science as cross-curricular class topics, a separate subject, whole-school topics, 'spur of the moment' investigations? Each system provides you with different problems to solve and each has different resource implications.

The whole-school topic, for example, will place a great strain on resources as it is likely that most classes will want the same equipment. At the same time, a large proportion of the school's resources will lie around gathering dust. If science is to be based to any degree on instant investigations, then each classroom needs its own basic science equipment. With the best will in the world you will not be able to supply enough to meet all needs, probably not even most needs. As a general guide, all classes should have instant access to some simple science equipment, such as magnets, and in particular to apparatus which extends observation – lenses, binocular microscope, cassette recorder, meters of various kinds, including a thermometer.

If your school does decide on a whole-school topic, you will need to ensure that the various classes tackle it in such a way that they will require different resources at different times. All classes should have access to a few pieces of specialist science equipment, but it should be used appropriately.

This brings us to the two systems of resource organisation most easy to cope with: class topics and separate science. In many ways, the organisation for both is similar. Both should be planned out for a given period with a range of expected activities or content areas and so both should have a limited resource requirement. If more than one or two classes intend to cover the same topic, or the content area is a little vague, then the problems mentioned above still apply.

I prefer to take a look at the topics that classes intend to cover in order to avoid clashes wherever possible. It also allows me to suggest activities or resources that the class teacher may be unaware of. The necessary resources for the topic can then be boxed up accordingly. If your school is particularly well resourced you may even wish to consider maintaining ready-prepared topic boxes for all or some of the most popular topics, though the problem with a topic box that someone else has made up is that it may not correspond to the way you want to cover the topic and will lead to an inefficient allocation.

Most commercial schemes and ideas books list the equipment required for each activity and class teachers should use these as a basis when planning their topics – as a basis, *not* as a straightjacket, since the school may well not have the exact requirements and the teacher, for that matter, may not need them. You need to recognise which items will be supplied from the science store and which ones you will need to find. It is also important to learn how to cope with alternatives if the piece you require is in use, or simply learn how to share.

I try to match what teachers want with what they actually need and usually this means giving them more rather than less. I also try to advise on where to go next with their class to build on the activities the children have taken part in, and the experience they have of various pieces of equipment.

Storage

The geography of the school is a very important factor when deciding where a central science store should go and, prior to that, if it is desirable. I would look first at how resources are stored for other curricular areas. If there is one system for maths, language or art, try to do the same with science. At least then it will be familiar to the staff and will be workable, as long as the existing system is.

- How accessible do you want to make it?
- To whom do you want to make it accessible?
- Will it be locked?
- Who will be responsible for it?
- Will children be able to collect certain items?

Central storage

I cannot answer these questions for your school, I can only make suggestions. First, if at all possible there should be a central storage area to which all teachers have free access. It should be co-ordinated and organised by one person or a small group. There should be a topic box and a small store of scientific equipment in classrooms to which children have access, assuming responsibility and safety of the children is accounted for.

Child access to class resources is very important for their scientific development. They will need to be able to choose their own equipment in order to demonstrate many desirable skills and attitudes, both within the science curriculum and well beyond it. The freedom of access to certain equipment will necessitate the children being aware of safe working practices and equipment usage, as well as the need to put things back where they find them.

Schools will need to think long and hard about the positioning of any central storage area, especially where there is a split site or mobile accommodation, to ensure that there is equal access. Many schools lack storage areas that can be made secure and safe for children. As a minimum you should have a lockable wardrobe in which to keep kitchen chemicals and glass or other fragile equipment.

Once in the store room there is no 'best' way to organise it. Just try to keep it fairly straightforward. I put all the optical equipment in one drawer or shelf, all the construction kits in another, film strips somewhere else together, and so on. Of course, if you have the equipment and a stable set of topics for your curriculum, you could always organise it by topics. Above all, keep it simple.

Classroom storage

If storage in schools is a problem, storage in classrooms is often non-existent! The bulk of your science equipment, mainly the expendables, much of the materials for investigation and 'junk' will need to be kept in the classroom. This could result in a disastrous mess if it is not managed effectively. Large corrugated cardboard boxes are useful, especially boxes of similar dimensions. For basic 'junk' materials they can be placed under tables or benches and hidden by drapes. For smaller, more organised storage, they can be turned on their sides, stacked between existing storage units and, covered with a coat of gloss paint or wallpaper, they look reasonably presentable and can be quite sturdy. Each of the boxes can then be subdivided with individual pull-out containers. (I use plastic four-pint milk containers which have had the quarter with the spout cut away). These can hold items such as rubber bands, cotton reels and tin foil separately.

If you are suffering from particular financial hardship, cast-off corrugated cardboard boxes can also be used as storage for topic loans from the central store. The best type of container is the stackable; for example, tomato or lettuce boxes can be picked up from either a market place or supermarket.

Another big handicap your school may have to face is trying to run a science curriculum with extremely limited resources. You can try to overcome the problem by structuring your topics to spread the demand for items in short supply. You may try to borrow the items as you need them from your local science centre (unlikely), organise 'swaps' with other local schools (risky), or face the fact that you need the item(s) and will have to make cuts elsewhere to get them. In the long run, only the first and last ideas are viable, which is why it pays to know what you've got.

Setting up a system

At this point, let's just recap on the stages we have gone through so far.

1 Perform an inventory of all science resources, including class-based, school and external ones. Note down all relevant resources and their present locations and, if possible, expose any obvious deficiencies that need to be dealt with urgently.

2 Where possible, allocate a set of basic equipment to each class to reinforce the children's ability to apply scientific skills in the normal classroom situation. Suggest types of materials and junk that would be useful for classes to start to collect. Give advice and assistance on the storage of these items.

3 In conjunction with the school's senior management or a staff meeting, decide on a central storage point for the remainder of the resources. Also, determine whether there will need to be changes to the teaching pattern for science in order to make the best use of existing resources.

So, there should be some equipment in classes and some in a central storage area, and someone should have some idea of where everything is. A reasonable start, but unless some form of organisation is involved the resources won't last out the topic.

After all of the equipment is neatly stored, there are four main aspects to maintaining your resources:

- Cataloguing (know what you've got);
- Booking-out procedures (know who's got it);
- Workshops (know what to do with it);
- Safety (know how to use it).

Cataloguing

Once you have made an inventory it is a fairly straightforward task to record the extra resources you may obtain, a slightly more difficult one to keep track of the items that go missing! If you have enough equipment to maintain topic boxes you will need to check them periodically to ensure that no 'borrowing' has taken place and that the box is usable. It reduces the amount of material storage required in classrooms but does put quite a strain on the organiser. You might well find it better to make up topic boxes of equipment as the need arises and negotiate the contents of them with the user. This is a task which can be delegated to children – they often enjoy it and are extremely thorough.

Booking-out system

To make sure that you are aware of items that go missing or the degree to which expendables are consumed you will need an efficient booking-out system. A good system is one that takes little effort to set up and minimal effort to operate. I have seen systems where the co-ordinator has made up library cards for each and every item and put pockets on the wall for teachers to drop their tickets into as they borrow something. Unfortunately, the tickets are not put into the pockets, are lost, or are not put back with the item on return. I have also seen systems where teachers are required to book out their resource requirements in advance. This normally leads to a very neat and full central store but a lot of frustrated teachers and disheartened children.

One system that seems to work is also probably one of the simplest: 'If before your topic or particular activity takes place you let me know what you need then I'll make up a box for you. If you want something immediately then go to the store, get it, and quickly jot down who you are and what it is you've taken. Or if you are in a hurry tell me or send a note to my class when you get time.' All this requires in terms of organisation is a well set out store, which is well labelled, and a clipboard or book with a pencil attached in some obvious position.

Another system that works is a row of hooks with 'class cards' on them. As the teacher takes a box, they put a class card in its place. You can then see at a glance who has what.

The by-word of any system you decide to use should be flexibility. You want maximum use out of the resources and you don't want to be tearing your hair (or anybody else's) out to see that it happens. It is well worth providing a box for 'breakages'. This brings into the open that everybody breaks something – even if it's only a flat battery – and gives you advance warning of shortages. Some breakages can be repaired, too.

With any system you apply you must ensure that everybody at least knows how it is supposed to work.

Workshops

Organise a tour of the store so that in those hectic moments everybody can find what they are looking for. A colour-coded plan placed in an obvious place (alongside the clipboard and pencil) will help, especially for those unfamiliar with the school – a supply teacher, for example, or the head!

You may build up a store of the most incredibly exciting and useful resources ever to be assembled in one school, but if nobody else knows about them or has no idea of their application or how to use them then they are useless. What you need to do is offer some practical guidance on how to make best use of what is available. I find that basing a workshop around a particular type of resource or equipment works best. Electricity is usually a good starting point, but others that have gone down well are:

● Construction kits;
● Using optical equipment;
● Light;
● Sound;
● Ourselves;
● Outside the school;
● Toys.

Obviously, there are many others – try to work to the strengths and interests of your staff. If you don't feel that you have the knowledge or confidence to do this yourself, look to your list of human resources and get an adviser or advisory teacher in to help.

Safety

The safety aspect relies heavily on the confidence, experience and basic common sense of the individual teacher, but the resource organiser can make a positive contribution by ensuring that no dangerous items are kept and by offering advice on how to use the resources. (The booklet *Be safe*, produced by the Association for Science Education, is invaluable – see the Resource List on page 72 for details.)

You should always aim to plan your curriculum with due regard for your resources, while at the same time planning your resource acquisition with due regard to your curriculum. When storing resources keep them ordered, organised and, above all, well used.

❸ Managing Science in the Whole Curriculum

According to the National Curriculum Council (1989), 'Attainment targets and programmes of study are the bricks with which the new curriculum must be built. Cross-curricular strategies bond these bricks into a cohesive structure.' In other words, the National Curriculum reflects good primary practice. Although a curriculum has been defined, it still remains for teachers to make the links between curriculum areas. Di Stead's chapter shows that topic work represents an excellent way in which to do this. After all, the real world is not divided into subject areas, but presents us with phenomena and problems that cross subject boundaries.

TOPIC : WATER

YO HO HO AND A BOTTLE OF RUM....

What is a topic?

Many teachers would answer this question by saying that a topic 'cuts across subject boundaries' and, of course, this is true. Most of us learned to write topic webs during our training, starting with a spider of an idea, spinning lines out to curriculum areas and then building links between these radial lines. Topic work is much more than this – it provides the opportunity for a different and distinctive approach to teaching and learning.

The organisation of children's activities can be liberated by a topic approach: tasks can be organised around the investigation of problems or the use of resources so that children can learn by finding out. Topics are flexible – they can be organised around individual, pair or group learning. The activities can be initiated by the children as well as the teacher – there is room for you to support children seeking their own path. Consequently, children can have a greater sense of ownership of their learning. Topics are often based on the local environment so that the school integrates with its community. At its best, topic work allows children to develop their own ideas and to grow as individuals.

Topic-based science

Learning in science can provide some of the most exciting experiences in a school. Motivating children to do science shouldn't be a problem – science teaching should tap children's natural curiosity. If you give a group of children three buckets of water, some plastic pipes of varying lengths and widths, and three different sized syringes and ask them which 'hosepipe' will squirt furthest, the problem is not one of motivation, but of maintaining some form of control!

Such experiences are powerful and direct and are the source of heated discussions where children talk about their experiences, struggle to find new vocabulary to communicate their ideas, and have personal views which are the basis for writing and drawing. The attainment targets in all three core subjects are often best met through channelling motivations that are generated by science experiments.

Science 5–16 (HMSO, 1985) suggested that science should be related to other areas of the curriculum and this is now the generally received view of good practice. Not only does the science curriculum have a great deal to offer in terms of exciting first-hand experiences, it also benefits from an integrated approach. The best teaching of science in an integrated topic results in sounder science learning where effective links are developed between science and the real world.

For example, following a visit to a fire museum, experiments with syringes and primitive pumps, a visit to a fire station and viewing films of fire fighters' movements, a group of children developed a dance based on their observations. The dance illustrated the operation of old and new types of squirts and pumps, and made a direct connection between how various pumps work, their effects and how pumps are used by fire fighters.

Choosing a topic

Choosing a topic within which to incorporate science is not always as easy as it might seem. An over-hasty choice of topic which appears to be appropriate at first sight can make life difficult in the long run. Broad topics can be tempting because it is easy to find lots of interesting areas that appear to fit in with science. However, you can easily end up juggling too many subjects within the topic, which is confusing for both you and the children.

Energy, for example, is a popular choice of topic which enables a large number of scientific subjects to be examined (e.g. human energy, electrical energy, nuclear energy, wind and solar energy, chemical, petrol and food energy, clockwork energy). The danger is that a programme could deal with each of these subjects in a superficial way and children would have difficulty making connections between this disparate range of subjects.

An alternative approach might be to consider the sun and its effects on the world, where children could experiment with the range of the sun's effects such as bleaching, heating, electrical impulse (as seen in a photo-electric cell) and lighting. The advantage of this approach would be that the children could gain understanding of the complexity of the sun's effects, and there would be greater opportunity for building progression and links into a series of experiments.

The HMI report (1989) gives two examples of topics that teachers choose because they appear to provide a good basis for science. 'Transport' and 'Communication' were both selected for their relevance to science but caused problems for teachers because the broad front of work quickly became unmanageable. A more focused topic may not seem so immediately appealing but it will be much easier to manage effectively.

Another pitfall is choosing a scientific subject area without considering the processes which are appropriate for learning science. Parachutes may provide a 'scientific' topic, but it is only when children make parachutes and devise a 'fair test' for them that the appropriate *processes* are involved in their learning.

It is easy to select a topic for its content and end up teaching it in a prescriptive way. This is particularly the case when difficult scientific concepts are involved. When choosing a topic, always bear in mind the processes and skills required in AT 1: raising questions, hypothesising, organising, designing experiments, experimenting, measuring, analysing results and drawing conclusions. The topic should be chosen for the processes and skills that it offers as much as for the content involved.

In the light of this, what do you do in science teaching?

- Look at your current topics.
- Which attainment targets in science are covered?
- Are there any gaps in attainment targets that need to be covered outside these topics?

Now look at the topic that you are about to teach. Brainstorm the topic and write your topic web or flow diagram in your usual way.

- What specific science content is there in your web?
- Write down the activities that you expect the children to be involved in.
- List visits, special equipment, visitors.
- Carefully examine these activities alongside the relevant National Curriculum attainment targets (look for specific scientific skills, personal qualities and content).
- How are these activities best ordered?
- Are you making any assumptions about what knowledge or skills the children will bring with them to the activities? Are these assumptions justified?
- Consider where any gaps in coverage of the attainment targets might lie and how these are best addressed (by other topics or by separate science learning).

Remember that you are not just trying to give the children the basics, breadth and continuity. You are also giving them a wealth of experience, the means with which to build their own understanding of how the world works and how science fits into that world. Above all, you are helping the children to develop questioning attitudes and open-mindedness in their approach to solving problems.

Co-ordination of topics

Continuity and progression are key factors to be considered when selecting topics. It is unfair and unhelpful to expect the teacher dealing with a class at the end of a key stage to 'mop up' a number of areas which are not related except for the fact that they have not been covered. Topics must obviously ensure coverage of the National Curriculum, but what if a topic does not fit well with the science curriculum yet deals with other aspects of the National Curriculum? In such cases, far better learning will result from avoiding the temptation to squeeze science into an inappropriate topic. Concentrate on science in the National Curriculum in a later topic.

It makes a great deal of sense to plan topics co-operatively as a school staff. The National Curriculum demands a carefully balanced approach and this is best not left to chance. It need not involve formal meetings, where staff agree a programme of integrated topics – an informal discussion in the staff room can be just as effective. However, it is crucial that sufficiently detailed records are kept to ensure coverage and ensure progression, particularly in relation to the processes required by the National Curriculum. An example of one school's policy on topics can be found in *Teaching and learning primary science* (Harlen, Harper and Row, 1988).

Resources

The key to an exciting and effective topic is the resources available to make the activities happen: it may be something exotic or unusual, it may be as prosaic as a sunny day. Whatever resources are available must be analysed in terms of the children's learning. For example, in a topic on 'Fire', I had available a museum of early fire fighting equipment and a large modern fire station. The order in which these two resources were used was not haphazard: it was essential to visit the museum first because the exhibits illustrated the basic principles of fire fighting. The visit to the modern fire station made more sense to the children because they could see similar principles at work in the machinery, despite the additional sophisticated technology.

A topic on toys and playthings might sound unexciting, but when we were loaned a selection of toys made out of junk by Indian and African children the topic came to life and provided the resources for a great deal of science and technology learning.

The collector who brought a selection of old bicyles into school provided the opportunity for a simple but valuable experiment. The children measured how far one turn of the pedals carried each bicycle. This provided a practical introduction to the cogs, gearing and the efficiency of the bicycles as machines. The activity that made this topic such a success depended on the local bicycle collector – the resources in your locality will be different but no less valuable.

However, resources need not be a matter of fortunate chance. It is possible to plan your resource and organise it in advancee, like the 'burglar' who entered our school, leaving finger prints, hairs and footprints which a forensic scientist was able to help us understand and interpret.

While it is important to seize whatever opportunities present themselves, do make sure that the resources selected are not racially or gender biased – all children should have access to the activities involved in the topic. (More advice on resources is given in Chapter 2, Managing Your Resources.)

Planning the topic

Planning serves two important purposes:
- it helps you focus clearly on what you want the children to achieve;
- it gives you confidence that the appropriate elements of the National Curriculum will be addressed.

When planning a topic, consider first what activities you expect the children to be involved in. Having identified these, consider what content and which skills and processes are involved. Now it is possible to decide whether these are the content and the process that you wish to address.

It may be necessary to alter the activities in order to focus on the part of the National Curriculum that the topic covers. It is only when the content and process involved in each activity have been considered that a judgement about the suitability of the activity can be made.

It is also important to consider how each activity builds on its predecessors and prepares for future activities. The *least* effective topic work is where too much is attempted in terms of content, and consequently little is achieved in understanding and skill development. It is unreasonable to expect children to pick up scientific ideas and skills simply because they are studying a relevant topic. The specific content and skills required must be identified and planned for. You cannot call your topic 'Science' if the children are doing no more than observing scientific phenomena. Their observations must lead into investigations. For example, close observation and drawing of a bird's wing do not by themselves involve the child in science. You need to plan so that the child might begin to ask questions and wonder about the structure of feathers and wings, and about how these relate to flight.

Planning for a topic must therefore be specific and clear: the investment of time and effort into the planning stage will save your time and energy during the management and evaluation of learning. Just as important, good planning helps to raise the level of your confidence before, during and after the learning.

Using published schemes

Not everyone has a strong science background, and it can be overpowering and bewildering for a non-specialist to be faced with the task of identifying scientific concepts and skills. There is a danger that the less confident teacher will take the route of least resistance and adopt non-scientific approaches. Electricity is one example of a topic where teachers can be tempted to bypass the science in the topic: children are sent off to find out about electricity from books and end up knowing a great deal about the regional distribution of electricity but nothing about how to handle wires and bulbs, make circuits and break them. The published primary science schemes are expensive but provide excellent support for the less confident teacher tackling science. Most schemes fall into one of two categories:

- schemes written around topics commonly used in primary schools – for example, 'Houses', 'Fabrics' or 'Festivals';
- schemes organised around science areas – for example, 'Energy and forces', 'Electricity' or 'Sound'.

The best schemes include an index which enables the teacher to select particular elements suitable to the attainment level of the children and the topic. Schemes which are based around topic areas usually have an index which indicates science activities suitable for each of the attainment levels.

Schemes based on science areas usually have similar check lists. It is worth looking closely at the range of schemes on the market before investing precious resources and, in particular, checking that the indexing system will help the most unsure teacher. Look for flexibility, the potential to incorporate your existing resources and adaptability to your own situation.

Fitting science in

Some areas of the science National Curriculum do not fit in comfortably with topic work. There is no point feeling guilty about teaching an area of the science curriculum outside a topic: if magnetism will not fit into the topic (and it may not) then leave magnetism out and teach it separately. The benefits of integrating science into a topic are dependent on the relevance of the science content to the topic. A contorted science subject which has been forced into a topic is of no value. The most important criterion by which learning is to be judged is that the children come out with a positive attitude towards the work. An integrated approach may be the best one, but using a topic does not of itself guarantee that the learning of science will be of good quality.

In the final analysis it is the quality of experiences that the children have – whether they develop specific skills, above all whether they enjoy themselves and have a positive attitude towards science – that determines the quality of the learning.

❹ Managing Assessment in Science – An Overview

The introduction of a national assessment and reporting programme involving both standard assessment tasks (SATs) and teachers' own classroom-based assessments presents a number of problems. Paramount among these is the amount of classroom time which will be necessary. Others include making the assessments as effective as possible, and presenting the results of the assessments in ways which can be easily understood by colleagues within the same school and elsewhere.

Martin Skelton here sums up what assessment means. The issues he raises are placed clearly in context in the second part of the chapter, when science co-ordinator Corinne Murray describes how assessment was successfully tackled in her school. There is no doubt that, to paraphrase the TGAT report, assessment lies at the heart of the learning process, enabling us to know what we ought to be teaching and how effective our teaching and children's learning has been.

HAVE WE REACHED LEVEL 2 YET MISS ?

Perhaps it would be better if we stopped trying to pretend that the answer to all our assessment problems lies just around the corner. It doesn't. The annual assessment of 30 children or so, in a number of different subject areas, in a way which will ensure that we obtain reliable and useful information, is always going to be difficult. This doesn't mean that we should abandon it. Precisely because assessment can provide us with some important and useful information, we need to constantly search for that 'best buy' – the assessment programme which provides us with the most useful information in the most effective way.

One of the ways in which such a programme will be developed is through an understanding of the issues involved.

How much assessment are we doing already?

One of the problems we face when contemplating the introduction of an organised assessment and reporting programme lies in assuming that it is all additional work. The truth is that a tremendous amount of assessment already takes place in every classroom every day. Within the past week, for example, you might have:

- made comments about a child's written work;
- talked with children about the fairness of a test they have constructed;
- observed and thought about children's attitudes towards science;
- overheard other colleagues talking in the staffroom about the science work of specific children;
- directed children towards one science activity rather than another because the one they were choosing was either too easy or too difficult;
- chosen models, investigations, drawings or written work for classroom display.

Each of these activities contains an element of assessment, defined by TGAT as 'a general term enhancing all methods customarily used to appraise performance of an individual or group'. The question is not, therefore, do we assess or don't we? We spend a great deal of our time assessing each day. The important question is, how effective are these methods of assessment for giving us the information which will really help us to teach science effectively?

What are the purposes of assessment?

The assessments we use need to be appropriate to the purpose we have identified. We might want to assess children to help us plan our future work. Such assessments are usually called **'formative'**. The most obvious point about formative assessments is that our emphasis needs to be on finding out what children don't know, as opposed to what they do.

But formative assessment can involve more than simply deciding which of the National Curriculum statements of attainment have not been learnt or understood. Much of the effectiveness of our teaching is dependent upon both the way we present the learning experiences to children, and the understandings children bring with them from their day-to-day lives. The Children's Learning in Science project based at Leeds University found, for example, that a significant number of young children identified a fire as 'living' in the same way that we are living. Following conversations with the children, one of the mooted explanations for this was their continual exposure to advertisements which talked about 'coming home to a living fire'.

While this might seem banal or even cute, such day-to-day understandings will have an effect upon the way in which children deal with the scientific experiences we present to them. So, formative assessment is more than just finding out what children don't know; it can also be about finding out the things they do 'know' that are likely to have an effect upon their understanding of new experiences. Formative assessments are all those which affect our future actions; they are the ones we use most often.

Secondly, we might want to assess children for reporting purposes to parents, other colleagues, new schools or for the children themselves. Such assessments are usually called '**summative**' and depend on us identifying what children know.

Thirdly, we might want to assess children to find out why they have difficulties in learning in order to identify areas of specific need. This '**diagnostic**' assessment can be used at the completion of summative assessments, but can also be a part of our ongoing classroom work.

What types of assessment are there?

There are numerous types of assessment of which it is worth identifying two groups:
- informal . . . formal;
- norm-referenced . . . criterion-referenced.

Informal . . . formal group

Most of the classroom assessments we mentioned earlier would be classified as *informal*, not because they are conducted in an informal way but because the results of the assessments need to be interpreted by us with some degree of subjectivity. The assumption (although an often-challenged one) behind *formal* assessments is that the results have some status independent of the subjective interpretation of the teacher or anyone trying to interpret them.

Norm-referenced . . . criterion-referenced

Norm-referenced tests try to place children in some form of rank order compared to other children. Norm-referenced assessment ascribes a grade to children which is concerned less with how well children have done than with how they compare against the performance of a whole group. Under a norm-referenced system, therefore, a child who achieves three-quarters of a given number of tasks may be graded higher one year than the next, depending upon the performance of the other children.

Criterion-referenced tests, on the other hand, aim to report the achievements of a child irrespective of the performance of others. Within criterion-referenced testing it is possible for all children to score highly. A criterion-referenced test sets out simply to record what children know; the results of such tests are much more statements of absolute achievement than norm-referenced tests.

What is important about norm- and criterion-referenced testing and the National Curriculum is that TGAT came down very heavily in favour of criterion-referencing – that children's performance should be described in terms of their understandable achievements. For a long time, teachers have criticised some of the iniquities of norm-referencing. Given some effective work by SEAC, the new assessment system should address some of those criticisms.

Are there particularly important skills in assessment?

The quality of *formal* assessments is dependent upon the skills of the test constructor to ask the appropriate question in such a way as to enable the necessary information to be revealed. Most teachers who have marked standardised tests will be able to remember the occasions when a child's answer to a question had to be marked wrong, even though it seemed a perfectly reasonable response to the question asked.

As far as we can tell, most of the assessments within the National Curriculum will be criterion-referenced, based upon teachers' understandings of children's performances in the classroom, or in specific assessment activities constructed to be almost identical to everyday classroom activities.

Observation and moderation

The first skill we need to practise, therefore, is the skill of **observation**. Accurate observation is much more difficult than it seems. As observers we need to:

- know precisely what it is we are looking for;
- go carefully to limit the extent to which our own prejudices affect what we see;
- find ways of checking that our observations were accurate.

The second skill we need to practise is that of **moderation**. Within assessment, *moderation* means the ironing out of differences between interpretations of events.

If much of the assessment in science is going to be criterion-referenced, then our ability to report fairly about the performances of different children depends on our shared understandings of what each statement of attainment means and what performance is required to have achieved it.

This is something which the National Curriculum still hasn't faced up to, although the programmes of study and non-statutory guidance have tried to help. There is still some way to go. To take just one example, AT 1 Level 3 asks that children should 'distinguish between a fair and an unfair test', one of the crucial scientific skills. But what does this mean as a Level 3 activity? Are children to distinguish fair and unfair tests every time before they can be said to have achieved success? Are they supposed to identify every variable in the tests, or just some? What performance counts as achieving this criterion? Does Johnnie qualify simply because he continually says, 'That's not fair!' (even though he may be right) or is more evidence and explanation required?

It is the process of moderation which helps to resolve these issues and which narrows down the possibility of error or unfair assessments. On the whole, teachers prefer criterion-referenced assessment because it is fairer to children. But its fairness, accuracy and usefulness is going to depend on our willingness to moderate some agreement among ourselves and our ability to observe what is really happening.

Are there any other key issues?

All teachers have their favourite key issues about assessment and this section could fill the whole book. I am concerned, though, about two further aspects of the assessment process. Let's just review the arguments to date.

Assessment provides us with important information. Assessment is a complex process. Criterion-referenced assessment, while fairer, demands greater skills on the part of the teacher than norm-referenced assessment. At any given time in your classroom you will be able to identify an enormous range of possible criteria to be assessed. You are probably teaching a class of between 25 and 34 children. You do have other things to do in addition to assessment.

Within that complexity we need a strategy to carry out assessments as effectively as possible. It is my view that we need to **focus** our assessments quite rigorously. Focusing means:

- not trying to assess too many things at once;
- not trying to assess too many children at once;
- spreading out the assessments over a reasonable time span;
- being clear about exactly what we are doing.

Effective assessment depends on our ability to:

- go into classrooms with a clear idea of what each of the criteria means;
- know what we need to see happen to suggest that the criteria have been met;
- know in our own minds about which activities we need to give our attention to;
- spend a reasonable amount of time with a group of children to enable us to be sure of our judgements.

We will only achieve this by focusing our attention. Apart from producing a nervous breakdown, going into a classroom with the intention of assessing all the children across all criteria as and when they happen will only produce an assessment which is hardly worth the effort.

The second issue simply responds to the question, 'What is this all about?' We know that assessment in science will help us plan our work, report to colleagues and parents and so on, but we need to ask, 'Why are we doing science anyway?'

I cannot believe that science has been introduced into the National Curriculum simply so that each child ends up with a checklist of ticks and crosses against all of the statements of attainment. I'd like to hope that children leave the primary school excited by science, understanding its relevance to their own lives, and aware of the part it can play in deepening our understanding of a whole range of issues. None of this is remotely guaranteed by the National Curriculum. Indeed, it is quite likely that some children might be able to achieve all that is required of them and still have a profound dislike of the subject. This would make a nonsense of the whole endeavour.

We need, therefore, to be constantly aware of children's attitudes and motivations as well as their absolute performance against certain criteria. We need to involve them in self-assessment procedures and in letting us know what they think, if all the work is going to be worthwhile. The implications for assessment are clear. As soon as we have come to terms with the introduction of SATs, with our own classroom-based assessment and with reporting, we need to think very quickly about introducing some form of **profiling** which presents us with a much broader picture of a child's scientific attitudes and motivations than a simple tick-sheet can hope to do.

– A Case Study

Sandylands County Primary is an urban school in Morecambe, Lancashire, where staff have devised their own system for assessing science skills and processes throughout the entire primary age-range. Corinne Murray, science co-ordinator, explains the reasons behind their approach.

Why assess?

When we embarked on our assessment programme we had very little idea of how we were going to set about it, but we were very clear about why. We felt that it was not enough to rely on some overall assessment of a child's general educational ability, but should obtain a more accurate and specific picture of each child's performance in various skills and processes.

We felt it was important to assess our children's performance in science for a number of reasons.

● To help us plan work at a more appropriate level for our children, either individually or in groups, within each class. In other words, to assist the process of matching the degree of challenge to the ability of the child. Unless that ability is known, the match cannot be appropriate.

● To become more aware of those children who may need extra help with science activities generally, or with specific scientific skills and processes. Children could otherwise slip through the net, their difficulties unnoticed, and we could be missing the opportunity to help them.

● To provide a more accurate picture of a child's abilities for the information of others (e.g. the child's next teacher, parents, headteacher, science co-ordinator, special needs co-ordinator and appropriate outside agencies). We were concerned that our school records were much more detailed and accurate for some areas of the curriculum – particularly language and mathematics – than for others and were, therefore, more useful for some curricular areas than for others.

We were able to report, for instance, on various aspects of each child's abilities in speaking, listening, reading and writing to a specific degree, such as the quality of vocabulary used in descriptive prose. But when it came to the science records our comments were general and less informative – 'Jane likes science and shows great interest' – which doesn't really help us to ensure that Jane is actually learning something when engaged in science activities at school.

● To help us to 'look and really see', not just assume we know each child's abilities. It was tempting to think that a child who showed a particular ability in some – or even most – curricular areas could automatically be assumed to have the same level of ability in the various scientific skills and processes. We thought it was important to base our records on facts not assumptions.

● To provide an overall picture of the class's levels of science abilities and so aid our future planning for the class. This type of information would also be vital for curriculum co-ordinators and the headteacher, in order for them to keep an informed overview of the progression of science skills and processes throughout the school.

● To address the assessment requirements of the National Curriculum.

What form should the assessment take?

Before we could decide on the way we were going to try to assess and record our children's performances we decided to lay down some preconditions.

1 It should satisfy all the reasons outlined above for wanting to assess children in the first place.
2 The recording system should be easy to complete.
3 It should be easy to read and understand.
4 It should be unambiguous.
5 It should be informative.
6 It should relate directly to the policy we had agreed in our school policy document.

The school policy document

Our document was completed – though open to periodic review – nearly two years before the final National Curriculum documents were available to schools and it reflected the relative importance our staff attached to a whole spectrum of scientific skills, processes, knowledge and conceptual understanding.

We also addressed the issues of balance, progression, cross-curricular activities and planning, classroom management, school management and organisation, resources, teacher planning and recording, and children's recording. However, the main emphasis of our policy focused on the development of scientific skills and abilities throughout the seven years of primary education.

To this end, we devised activity lists for each of the three age-groups we identified at school, that is infants (ages 4 to 7), middle school (ages 7 to 9) and upper school (ages 9 to 11).

Communication   Infants
Describing verbally, pictorially
Observation   Infants

Activity List   Infants
Investigating –
Asking questions
Predicting outcomes
Drawing conclusions
Relating conclusions to the original problem
New investigations following these conclusions
Collecting materials for observing and investigation

Academic Year	Class	Activity List: Infants

1 Investigation Activities
- Raising questions.
- Proposing and arranging investigations.
- Introduction to fair testing.
- Persevering with experiments.
- Collecting materials for observation and investigation.
- Predicting outcomes.
- Drawing conclusions.
- Relating conclusions to original problems.
- Designing new investigations in the light of these conclusions.

2 Observation and Discrimination Activities
- Observing and becoming aware of characteristics of living and non-living things.
- Listening and hearing.
- Being aware of similarities, differences, changes, patterns and sequences.
- Classifying and comparing.
- Estimating and measuring.
- Identifying relationships.
- Using observational apparatus appropriately.

3 Communication Activities
- Communicating and describing verbally, pictorially, physically and in writing.
- Organising information, making connections.
- Recording and interpreting data.
- Following instructions.
- Using appropriate terms associated with shape, size, length, weight, time, growth, cause and effect, volume and capacity.

4 Social Activities
- Participating and co-operating in groups.
- Sharing materials and time.
- Caring for materials and the immediate environment.

5 Manipulative Activities
- Handling materials.
- Carrying out experiments.
- Hand to eye co-ordination.
- Awareness of safety, accuracy and tidiness considerations.

Academic Year	Class	Activity List: Middle School

1 Investigation Activities
- Identifying problems.
- Making hypotheses.
- Proposing, arranging and planning related to investigations (including equipment).
- Predicting outcomes.
- Designing fair tests.
- Carrying out investigations methodically, responsibly and fairly.
- Drawing conclusions.
- Relating conclusions to original problem, hypothesis and prediction.
- Using information from a variety of sources.

2 Observation and Discrimination Activities
- Estimating and measuring.
- Selecting relevant observations.
- Pointing out similarities, differences, changes in patterns and sequences.
- Classifying and comparing according to criteria.
- Identifying relationships.
- Using observational apparatus appropriately.

3 Communication Activities
- Communicating and describing verbally, graphically, pictorially, and in writing.
- Recording and interpreting data in above forms, honestly.
- Planning practical activities.
- Using quantitative records of data.
- Using appropriate terms, related to above.

4 Social Activities
- Showing respect, co-operation and participation in groups.
- Taking appropriate share of initiative and responsibility in groups.
- Showing respect and concern for living and non-living things.
- Carrying out group investigations.
- Showing respect for others' point of view.
- Caring for immediate environment.

5 Manipulative Activities
- Co-ordinating hands and senses.
- Working safely, accurately and tidily.
- Handling equipment appropriately.

Academic Year	Class	Activity List: Upper School

1 Investigation Activities
- Identifying problems.
- Making hypotheses.
- Designing and planning experiments, materials, apparatus.
- Making predictions.
- Designing and carrying out fair tests.
- Controlling variables.
- Re-testing to ensure fair result (use of average).
- Drawing valid conclusions and generalisations on basis of results.
- Relating conclusions to original problem hypothesis, prediction.
- Recognising limitations of conclusions.
- Designing better or more appropriate investigations in the light of findings.
- Using information from a variety of sources.

2 Observation and Discrimination Activities
- Using estimates and measurements appropriately and accurately.
- Recognising need for observations relevant to time and space.
- Observing selectively and in appropriate detail.
- Using observational apparatus appropriately.
- Classifying accurately, according to criteria.

3 Communication Activities
- Selecting and using most appropriate form of communication for recording and interpreting.
- Planning investigations sequentially.
- Selecting and keeping appropriate quantitative records.
- Reasoning critically.
- Using appropriate terms relating to energy, materials, living and non-living things, technology.

4 Social Activities
- Participating and co-operating in an open-minded manner in groups and individually.
- Showing respect for others' point of view.
- Becoming self-critical.
- Working independently and interdependently.
- Caring for immediate environment.
- Developing respect and concern for the wider environment.

5 Manipulative Activities
- Manipulating appropriate equipment with concern for safety, efficiency and tidiness.
- Preparing and making own apparatus when appropriate.

Each activity list consisted of a collection of skills and processes, divided into five categories:

- Investigation activities;
- Observation and discrimination activities;
- Communication activities;
- Social activities;
- Manipulative activities.

(These categories were chosen based on literature published by Lancashire Education Authority.)

At the time of formulating the policy document, every skill or process listed under these categories for each age-group had been discussed, changed, moved, or reworded until the finished lists were acceptable to all members of staff. We intended to use these to ensure a balance of skills and processes throughout a year's science topics, and to ensure progression of skills and processes throughout a child's science education from age 4 to 11.

We did not produce similar lists for each age-group covering scientific knowledge and conceptual understanding, preferring instead to adopt a concept list produced by the Lancashire Primary Science Curriculum Development Group. This listed the concept statements we intended to develop throughout the primary age-range and which we would try to ensure most children had grasped before they moved on to secondary education.

Of course, this implied an emphasis on the acquisition of skills and processes involved in scientific activities, rather than simply the assimilation of scientific knowledge, while acknowledging the importance of knowledge and understanding as a result of scientific explorations.

This emphasis reflected our view that the overall characteristic of good primary science is children finding things out for themselves in a challenging environment.

Obviously, whatever form our assessment was to take, it had to reflect this view and involve, in some way, the activity lists which were to be the backbone of our school policy.

Our options for assessing and recording

As science co-ordinator, my role was to put forward some options for assessing and recording for the staff to discuss, add to and amend until we found a method we were happy to try.

1 Written comments (sentences) referenced to our activity lists, based upon our own subjective observations and discussions with each child as part of normal classroom science activities.

Although this option would fulfil conditions 1, 3, 4, 5 and 6 above (see page 46), it would not be easy to complete in a busy classroom with 36 children! Furthermore, a teacher who is trying to write notes on a child's performance, referenced to a predetermined list of activities, would not be able to act efficiently as a teacher at the same time.

2 As above, plus subjective comments about the children's conceptual understanding and assimilation of scientific knowledge.

This really did seem like a Herculean task! Especially as it would also require forming separate concept lists for each of the three age-groups. We really felt that, hard working, keen and conscientious as we were, this was going over the top!

3 The use of tests, such as those used by the Assessment of Performance Unit (APU).

Although this was the only option which would give us an objective assessment and some form of standardised score, we rejected it for a number of reasons.

- We were unsure of the availability of such material for all age-groups.
- We did not want to subject the children to the pressure of a 'test', which may advantage some and disadvantage others. Besides, it seemed unsuitable for the youngest members of our school unless conducted in a very teacher-intensive, time-consuming and therefore impractical manner.
- Above all, we were concerned that our assessment should be a meaningful part of our normal classroom science activities. We wanted to find out what children could and did do during science activities at school in the context of a relevant and meaningful area of content.

4 The use of some kind of scoring system — probably a 3 or 5 point scale — referenced directly to our activity lists, based upon our own subjective observations and discussions with each child during normal classroom science activities.

We felt that this option would fulfil all the conditions we had laid down, but accepted that a scoring system has some definite disadvantages.

Primarily, we acknowledged that a child cannot be described in terms of performance, ability or anything else by the use of a number. But we did feel that this method would be relatively easy to complete and record and that it was a way of assessing the children's performance directly in relation to a specific skill or process.

This decision also meant that we would not be assessing the acquisition of knowledge or conceptual understanding at all at this point.

Reaching a decision

Once we had decided to adopt a subjective scoring system, directly referenced to our activity lists, we then made the following decisions about the way we would carry out the assessment.

1 We would pilot our assessment for two terms before deciding on whether or not to include it in our school policy and the final form it should take.

2 We would not set any specific tasks or tests for assessment purposes. Rather, we would simply assess as part of our normal observations, discussions and challenges of children in science activities.

3 We would use a 3 point rather than a 5 point scale where:

- 1 = independent in a skill or process;
- 2 = needs some support in a skill or process;
- 3 = unable to perform a skill or process.

We felt that this would lead to less ambiguity than

a 5 point scale, where one teacher's '3' might be another's '4'. We also thought that with a 5 point scale we might be less inclined to use the extremes of the scale.

In practice, those teachers who felt constrained by the limits of a 3 point scale occasionally used a score such as 1/2 or 2/3. So, even a 3 point scale proved to be flexible enough to cope with most eventualities.

4 We would try to use as little paper as possible for recording purposes.

This wasn't for any reasons of environmental conservation, more for practical ones – if each curriculum co-ordinator asked each class teacher to complete one sheet of paper to assess each pupil in the class, we would soon all be pushing filing cabinets around school. Besides, who would ever have the time or inclination to read all those pieces of paper?

We decided to try to collect all the assessments for one class on one, or at the very most two, sides of A4.

5 For the purposes of piloting the assessment project, teachers could choose which skills or processes they wished to assess, but we would concentrate on the first two categories of our activity lists – investigation activities and observation and discrimination activities - as we felt that these two areas were less well represented in other curriculum areas.

Teachers of parallel age-groups would choose the same skills to assess.

With the introduction of the National Curriculum, we considered replacing our activity lists with the statements of attainment found in AT 1 and using the other attainment targets to assess the whole spectrum of primary science. However, many of the statements of attainment are simply not specific enough to be used as a reference point for assessment. For example, AT 13 'Energy', Level 2, requires children to 'be able to describe how a toy with a simple mechanism which moves and stores energy works'. But at what point can one say that a child has reached this level of attainment? Is it enough for them to say, 'Because it has a battery', or do they need to go into detail about gears? And what is a 'simple mechanism'?

We spent a whole in-service day debating these issues and agreed that the best way for us to achieve some consensus about how to judge what a child has achieved was to link the criteria to the activity lists in our own policy document, so that we were working with skills and terminology with which we were all familiar. Other schools can only do the same, following their instincts to devise a system which works for them.

Putting it into practice

In practice, we were amazed at how simple and straightforward this assessment procedure seemed to be.

Teachers of parallel and similar age-groups found it beneficial to meet regularly to discuss progress, practice and ensure parity. Most teachers found it possible to assess the whole class on at least five

ASSESSMENT OF SCIENCE

CLASS RECORD SHEET

Academic Year __1987-8__

Class __5 MURRAY__

For each skill or process, record the appropriate number:

1 - Independent in that skill or process
2 - Needs some support
3 - Unable to perform skill or process

Record the skill or process and the date of testing in the sloping columns.

NAMES OF PUPILS		Making Hypotheses 1b (Oct '87)	Planning investigation 1c (Oct '87)	Fair testing 1c (Oct '87)	Drawing Conclusions 1e (Dec '87)	Estimating 1g (Dec '87)	Selecting relevant observations 2a / 2b (Dec '87)							
Angela	BROWN													
Michael	CATHCART													
Sajid	DINA													
Wayne	GATES													
Emily	HARRISON													
Hayley	HEWSON													
Naveed	PATEL													
Sara	PHILIPS													

specific skills each term. Some managed considerably more.

Many teachers liked to reassess children at a later stage in the school year for the same skill, particularly if children had performed at a fairly low level at the initial assessment. The provision on the record sheet for the recording of the date of assessment made this action clear to readers. We found it most useful to assess the entire class for the same skills over as short a period as possible.

Sometimes we would assess everyone for just one skill over a very short period of time, and sometimes we would assess two or three skills for the same group at the same time and progress around the class from group to group in that way for a slightly longer period of time.

The completion of county pupil records and parents' reports was greatly eased and improved by having the record sheets at hand. Salient features of each child's performance as shown on the class record sheets could be written in sentence form on such reports – for instance, a score of '2' for one skill and '1' for another could be written as:

> Jane needs some help when drawing valid conclusions from her science explorations but is now able to design and plan her own experiments

In the classroom

To give an example of how assessment was actually carried out as part of normal science activities, I'll explain how I assessed the 9- and 10-year-olds in my class for just two of the skills involved in their science activities.

The science made up one aspect of a half-term topic on 'Transport'. Various aspects of the mechanics of transport were investigated throughout the half term, but one particular, open-ended challenge involved a few weeks of quite intensive investigation by the children at different levels, according to their ability and tenacity.

The challenge was three-fold:

```
TRANSPORT

1 - Can you find out how much force or energy
    is needed to pull or push this (toy) truck
    up this slope ?

2 - Can you make it harder - so that it needs
    MORE force or energy ?

3 - Can you make it easier - so that it needs
    LESS force or energy ?
```

I assessed the children on two of the skills and processes listed for this age-group:
- making predictions;
- making hypotheses.

Making predictions

I was not looking for a prediction which was scientifically correct – simply one which related to the challenge and could be tried out. Some children began with an attempt at a prediction, for example:

- 'It will be hard to pull it up the slope.'
- 'You could put bricks in.'
- 'Some slopes are harder or easier if they're made of different things, like wood or plastic or carpet.'

These types of statement were not quite predictions because they couldn't be proved or disproved as they were. The children often needed a little support to refine these into proper predictions, such as:

- 'It will be harder if the slope is steeper.'
- 'It will be harder with bricks in the back.'
- 'It will be easier on a plastic slope surface.'

Those children who couldn't progress from the first kind of statement to the second, even with support, would score a 3, although in this case everyone actually managed to score 1 or 2.

Making hypotheses

Again, I was not looking for scientific accuracy but a statement backed up by a reasoned theory which could be systematically tested.

A few children scored 1 for this skill as well. Examples of some of their hypotheses were:

- 'The steeper the slope, the more the energy that will be needed to travel up it.'
- 'The heavier the load, the greater the energy needed to climb up a slope.'
- 'The smoother the surface, the less the energy needed to travel up it.'

Many children needed support – by means of a discussion with me – to progress from their specific prediction to a more general hypothesis. These children scored 2.

Those children who were unable, even with considerable support, to make the jump from a prediction to a hypothesis, scored 3.

Using the results

As the topic progressed, I was able to extend my assessing to cover other skills and processes but, more important, I was able to keep referring to my records of children's performance to develop those skills in which my assessments had indicated some support was needed.

Thus, my class record sheet became a useful teaching tool for the rest of the academic year, and hopefully for the next teacher of the class.

Learning from our experience

At the end of our pilot period, we certainly felt much more experienced and confident about the assessment issue. We had learned how to carry out this form of assessment with as few problems and disruptions to our normal teaching as possible, while making it as meaningful and useful as we possibly could.

Here are a few of the lessons we learned.

1 Always use some form of reference for assessment which is specific, unambiguous and appropriate to the ages and abilities of the children concerned.

2 Don't try to assess more than three different activities or skills at the same time.

3 Try to assess the class for the same activities or skills over as short a period as possible in order to ensure uniformity of approach and to obtain parity of results.

4 Jot down your judgements **there and then**, not on reflection later that same day, at the weekend or half-term. This would be assumption, not accurate observation.

5 Use as many different methods as possible in order to make your judgements, e.g. discuss, observe, question, set mini-challenges within the task in which a child is engaged, in order that your decision is as informed as possible.

6 Where children usually work in groups, ensure that these groupings are changed regularly in order to ensure that you are carrying out a true and fair assessment of the individuals within each group.

Assessment really did open our eyes to the variety of levels of skills and abilities within our children, collectively and individually. Such readily obtained, relevant information has to be an asset for the teacher who is able then to act on it to improve the level of science activity in the classroom.

At the end of our pilot period we all agreed that assessment isn't something new and frightening, if tackled in a realistic and practical way. It is merely what we have been doing for years – noticing things about children, and learning and using what we notice to help them learn more.

Case study review

The staff at Sandylands County Primary have clearly done some tremendous work over the past two years and many of the issues they have already faced about assessment will be of concern to all schools in the next year or two. On the basis that there seems little point in re-inventing the wheel, is there anything which can be learned from this descriptive case study which might help schools beginning to face the problems already encountered by Sandylands?

First, of course, we need to notice the structured way in which the school went about its developmental work. Although such an approach takes a little longer, it almost always produces better, more thoughtful results. Second, we ought to notice the way in which the science work undertaken by the school both fed off and contributed to the positive atmosphere between colleagues at the school.

I'm going to suggest that the school might need to think carefully about the crucial aspect of moderation. In doing so, I am aware of the dangers of this, partly because Sandylands might have already begun to tackle this important stage and partly because the case study is unlikely to have provided all the information that exists. So my next paragraphs don't really refer to Sandylands at all; they are more concerned with using the case study as an example to draw out a couple of points.

Moderation is about 'the process of checking the comparability of different assessors' judgements of different groups of pupil' (TGAT Preface and Glossary 1988). If we don't moderate the assessments we make, then any comparability of those assessments is virtually meaningless. You will remember that Sandylands chose a three point scale initially: 1 meaning 'independent in a skill or process', 2 meaning 'needs some support in a skill or process' and 3 meaning 'unable to perform in skill or process'.

To assess the outcomes of children against the categories requires everyone who uses them to be very clear about exactly what they mean. Does 'independent in a skill or process' mean always independent, with never the need for help? If it does, how many '1s' would you score in your life? How much support is the 'some support in a skill' which might be needed? Some teachers at Sandylands clearly realised the problems and started to award scores of 1/2 or 2/3; in other words, some children were identified as needing some support but not as much as others.

The only sentence in this exciting case study which caused me any real concern was the one which read, 'In practice we were amazed at how simple and straightforward this assessment procedure seemed to be.' Moderation is not a simple activity but it is – along with teachers' well-developed skills – one of those activities which holds the key to effective criteria-referred assessment.

Any school determined enough to follow the Sandylands route should always be asking three questions:

1 Do we share an understanding of what the criteria mean?

2 Can we agree what the children have to do to satisfy these criteria?

3 How can we check that our observations of those outcomes amongst different children are as similar as possible?

When we can get as close as possible towards answering all three questions positively then some really effective assessment is likely to be taking place. I'm sure Sandylands are working at it right now.

⑤ *Managing Science in the School*

The headteacher's task in developing science in school is straightforward. Gather the staff together at 4 p.m. on a wet Friday afternoon and tell them that from the following Monday morning everyone must ensure that their classes are engaged in a structured, cohesive and coherent science programme which takes up a reasonable amount of time and is matched to each child's individual needs within a broad and balanced curriculum.

If only it were that easy! Mike Sullivan explains how it can be done.

Until recently, successive reports have singled out the teaching of science in primary schools as an activity which in many cases is done badly, if done at all. Mountains of print have appeared about the teaching of primary science and numerous projects undertaken, but actually getting teachers to engage children in science activities in classrooms has been an uphill task. Now teachers and schools have no option but to teach science to children. The National Curriculum is neither optional nor negotiable. We have to deliver the programmes of study, and we must be seen to be trying to achieve attainment targets, no matter how unrealistic some of these may appear to be.

Headteachers need to convince, enthuse and motivate staff so that good science takes place in classrooms. There are other important players in the game whose commitment also needs winning. Increased responsibilities devolved to governors mean that they need to understand the importance given to science in the curriculum. Parents, too, need to be well informed about what is taking place in school.

Setting the budget

As school governors take on more responsibilities for schools' budgets under arrangements for the local management of schools (LMS), it becomes increasingly important for them to make decisions based on evidence and fact, rather than on guesswork, intuition or prejudice. Governors need to be provided with options and recommendations at their meetings, rather than asked to endorse unilateral decisions reached by the staff or the headteacher. In law, governors are responsible for the curriculum and finances of a school, and the headteacher for a school's organisation and management. In setting a figure for materials and equipment under LMS and dividing that figure under expenditure headings, governors will need a clear idea of priorities, particularly in providing resources so that the school can effectively meet its statutory obligations to deliver the National Curriculum. It becomes increasingly risky to fail to consult governors at an early stage in decision-making on major curriculum issues, as without a full understanding of circumstances and needs governors may well reject staff decisions and impose their own priorities and time scales.

Curriculum leadership

One of the outcomes of the Education Reform Act may be the loss of the last vestiges of teacher autonomy within the classroom. No longer can the doors be shut so that teachers can teach their classes exactly what they want in the way that they want. The autonomy of the individual teacher has been replaced by the collegiate responsibility of the whole staff to deliver the National Curriculum in the most effective and efficient way possible. There must now be discussion and agreement about curriculum content, teaching method, pupil progress and record keeping, and there needs to be someone on the staff to lead and co-ordinate work in key curriculum areas. As late as March 1987, the HMI Primary School Staffing Survey showed that there were far fewer primary teachers with leadership responsibilities for science than there were for the other two core subjects of English and maths. The figures were:

- 16 700 primary teachers who were curriculum leaders for science;
- 20 700 with responsibility for English;
- 19 600 with responsibility for mathematics.

This means that there were well over 20 per cent more curriculum leaders in primary schools covering English or maths than there were covering science.

Even though rapid changes are taking place in organisation and management of schools, there are still likely to be significant numbers of schools without science co-ordinators, and a great number of schools where the science co-ordinators have been newly appointed.

Schools establishing a new post for science, and even those that have had science posts for some time, will be most likely reviewing the job specification of the post holder to reflect changes in the role as developments occur. Any review will probably have to not only identify what needs to be done, but also plot out ways in which the post holder can realistically meet those needs. Job descriptions normally cover both curriculum and interpersonal skills and should be public knowledge so that all staff are clear about each other's area of responsibility.

The job specification on the next page is not exhaustive, but it does identify some of the key areas which need to be covered by a science post holder.

A useful chapter on the role of the curriculum post holder is to be found in Campbell (1985).

The 1987 HMI Staffing Survey found that only 18 200 out of a primary teaching force of 159 600 (just over 11 per cent) had a GCE 'A' level in any science subject, so those schools looking to recruit a teacher with a formal scientific background are unlikely to be overwhelmed by applications from highly qualified science specialists. Although the lack of scientific background in the majority of primary teachers has implications for the sort of science that teachers feel confident to undertake in school and therefore for staff development and training, one must question whether a science qualification is any benefit in teaching primary science. Is it not more important to be a good primary practitioner?

Decision making

Offering an adequate range of activities and experiences that provide continuity and progression is a difficult task for any school. An option that is obviously worth considering is the purchase of an 'off the peg' commercial scheme. Unfortunately, unlike suits, publishers' schemes only come in one size. The scheme will fit where it touches – with any luck a school might find one which fits its philosophy and has enough elbow room to enable staff and children to feel comfortable with it. Commercial schemes have certain disadvantages.

Curriculum leader in science
JOB SPECIFICATION

Key Areas	Key Tasks	Implementation	Review
Curriculum	• Keep up to date with statutory requirements.	• Read and review all NCC and SEAC Documents relating to Science (own time).	
	• Keep up to date with other developments in science teaching.	• Read journals and books (own time) and attend courses (mix of own time/school time).	
	• Examine and modify school science scheme in the light of NC requirements and education authority curriculum statement.	• Consult with all other members of staff to include the setting of target dates for the introduction of the whole or parts of science scheme (own time).	At least annually.
	• Develop record keeping and monitoring systems.	• Consultation (own time).	Annually.
Resources	• Listing existing resources and checking on whether resources are in working condition.	• Gathering of resources for stock taking (own time).	Annually.
	• Identifying additional resources needed.	• Consult with others (in-service days).	Annually.
	• Organising, storing and access to resources.	• Consultation (own time).	Annually.
Professional Guidance	• Run staff workshops on science teaching.	• Consultation (in-service days).	
	• Encourage others to attend science courses.		
	• Establish and maintain links with support services.		

- They are expensive in cost of books and expendable pupil materials.
- They naturally cannot take into account and exploit the unique features of the school's local environment.
- Developments in other areas of the curriculum within the school cannot be predicted by a commercial scheme and so cross-curricular links are not always forged.

On the other hand, using a commercial scheme can give teachers with little formal background in science confidence in what they are about. Commercial schemes are written by experts and enthusiasts, aren't they? So they must be sound. Unfortunately, even examination by a semi-critical eye would reveal that some material for junior-aged children is not particularly child-centred but is watered down secondary science, and that for infants is the same content that has suffered yet a further dilution.

The answer is to choose a scheme with sufficient flexibility to enable the school to develop its own use of it and incorporate current materials.

Taking risks

Learning and teaching through exploration and discovery rather than by didactic 'talk and chalk' methods is in many ways unpredictable and risky. Creating and maintaining a climate within the school where intellectual risk taking is encouraged is very much the responsibility of the headteacher.

Exploration in science is a voyage of discovery for children. Even though the voyage is guided by the teacher, the results of children's activities can be unexpected and it may well be appropriate that children explore blind alleys before drawing conclusions. Obviously, the learning outcomes of some activities will be greater than others, and some activities will fail to reach the goals set. Peters and Austin, in their book *A passion for excellence*, say that:

> *"The best bosses — in school, hospital, factory are neither exclusively tough nor exclusively tender. They are both: tough on the values, tender in the support of people who dare to take a risk and try something new in support of these values."*

When things don't turn out quite as expected, or there is criticism, then the role of the head is to take responsibility and not to point the finger at others.

The quality of the success or failure of a school in any curriculum area is in the children's attitudes to that subject and the knowledge and skills that they have developed. The focus of the National Curriculum and SEAC documents is not entirely on knowledge and skills. Listed in *Science in the National Curriculum* as important attitudes and personal qualities in science education are:

- curiosity;
- respect for evidence;
- willingness to tolerate uncertainty;
- critical reflection;
- perseverance;
- creativity and inventiveness;
- open-mindedness;

- sensitivity to the living and non-living environment;
- co-operation with others.

Unfortunately, there is no reference to enjoyment or pleasure in learning or teaching. Technical competence without commitment or enthusiasm has all to do with training and nothing to do with education.

Managing tension

As a task of scientific data gathering I took the measure of the National Curriculum and SEAC documents that have arrived at school during a two year period. Putting them all on the scales, they weighed in at over seven kilograms – about ten times the weight of my copy of the complete works of Shakespeare!

The TGAT reports, including the three supplementary reports, weighed in at 750 g and 278 pages. The NCC Science Consultation Report weighed in at 600 g and 148 pages, the Science Group Interim Report (Nov 87) weighed in at 300 g and 82 pages and the ring binder alone (*Science in the National Curriculum* NCC (1989)) weighed in at 1 kg. Among its 179 pages are such mind-blowing gems as:

> "Subject to paragraph (3), the programmes of study described in this document and set out in column 2 of Schedule 2 to this Order are specified in relation to the key stages set out beside them in column 1 of that Schedule."

Attempting to get to grips with this bureaucratic nightmare of paper, dealing with the day-to-day management problems of running a class or school and maintaining some sort of family life, puts all in school under a great deal of personal pressure and stress.

Establishing priorities and producing a reasonable workload for staff is very much the task of the headteacher. The following list taken from *Science in the National Curriculum* describes the judgemental aspects of design and technology. The list will also serve as a useful guideline for tackling the National Curriculum.

- deciding what is worth doing and is achievable;
- generating and appraising possible solutions;
- reconciling conflicting demands;
- making decisions on the basis of imperfect information;
- achieving outcomes within constraints of time and cost.

Implementing new styles of management within the school should, in the long term, ease the burden but, in the initial stages, new styles of management will add to the load of tension within the school.

How do people know what's happening?

One of the main causes of tension is uncertainty caused by poor communication. The headteacher must ensure that teachers are fully informed and involved, yet not buried under mountains of paper or occupied with endless briefings at curriculum meetings. One way of keeping the paper down is to insist that school policy documents cover specific areas and that overlap is kept to the bare minimum (e.g. general policy documents on forecasts, marking of work, record keeping and resource management can cover all curriculum areas).

Meetings

An agreed agenda distributed well in advance of a curriculum meeting and an action column contained in the agenda can move the business of the meeting along at a good pace. Minutes of the meeting, not only including decisions made but identifying those people who have promised action and a time scale in which action is to take place, also reduces confusion and frustration.

Keeping an eye on things

At a time of radical change it is more important than ever for headteachers to keep a finger on the pulse of classroom activities. Traditionally, heads without full-time teaching commitments have spent time in classrooms relieving post holders to carry out other duties, covering for absences and providing general support. Wandering about the school gives heads a good idea of what is taking place but rarely have heads engaged in regular systematic observation of teachers or children. Headteachers see themselves as leaders and people of action rather than passive observers. Gathering hard data from observation of teaching and learning may well be a new task. One practical approach is to 'shadow' a child on a mixed teacher day and see what really happens.

It is unlikely that the curriculum audit that schools will be obliged to undertake will be as rigorous as initially planned, but schools will probably need to publish at least rough estimates of the percentage of time taken up under a number of curriculum headings. The head needs to be a regular observer in classrooms, not only to confirm that science is taking place, but that the tasks that children are undertaking are appropriate, and that the children are engaged in active rather than passive learning. Education authorities, too, have a responsibility to ensure that the National Curriculum is being properly followed in schools, and inspectors are now taking a much more systematic role in monitoring classroom activities. The head who is undertaking classroom monitoring is much better placed to confirm or challenge inspectors' observations.

What about parents?

Parents, too, need to know about new developments at school. The legal requirement that all parents should have written reports giving details of their children's progress in the National Curriculum isn't enough. Knowing that Jason or Charlene has reached Level 4 on ATs 1, 5, 7 and 9, and Level 5 on all other ATs, may be helpful to the minority of parents who actually fork out £6.95 to HMSO for the relevant document, but for most parents this kind of report will be virtually incomprehensible.

Now, more than ever, there is a need for schools to hold science evenings and workshops for parents so that developments can be discussed and ideas shared. It's a good idea if children are invited back to school one evening so that the parents can see the children actually at work. If this isn't possible, then a video of groups of children carrying out a variety of science tasks makes a good substitute.

Enticing busy parents to turn out in the evening can be difficult. One school invited parents to come to a science evening but laid down the condition that parents should bring a teapot. During the evening the parents were asked to find which was the best teapot. An open question which resulted in comparisons of volume, weight, heat retention and ease of pouring – a real exercise in the exploration of science. Talking *with* parents is a much more fruitful exercise than talking at them.

In-service training and school-based support

The shortage of staff in primary schools with any sort of scientific background presents a serious problem. The headteacher in drawing up the School Development Plan will need to take into account the ways in which in-service training can bring about learning through discovery in school. Centre-based courses provide a means of developing skills, but even more valuable is the involvement of science advisory teachers working in classrooms alongside teachers. There is nothing more useful in in-service development than sharing a teaching activity with someone who has expertise, has empathy and is working with your children and under the constraints with which you normally work.

Praise and encouragement

What is needed by any headteacher to develop successfully an investigative approach to science (or any other area of the curriculum) in school is the ability to identify key tasks, identify key people to perform those tasks, and to offer the maximum amount of support possible. Time and resources are important areas of support, but even more important is to give praise generously when things go well, and wholehearted encouragement at times when progress is slow.

Finally, a health warning for all those working in curriculum development – enthusiasm and energy are infectious and are spread by investigation and discovery learning.

Resource List

Books and References

Althouse, R *Investigating science with young children*, Cassell 1989.

Examines the processes involved in science; activities described in detail; examples of questions to ask with ideas on how to extend and challenge children's ideas.

Aspects of primary education: the teaching and learning of science, HMSO, London 1989.

Trends and issues which have influenced the standards of work in primary schools since the publication of the National Primary Survey in 1978. Particular attention is given to examples of good practice and to the implications for primary schools of the introduction of the National Curriculum.

Avon primary science working papers, available on a variety of topics, e.g. *An approach through problem solving* and *Science and technology from topics*. Copies can be purchased from Maths, Science and Technology Centre, Bishop Road, Bristol B57 8LS. Also Avon's *Primary science guidelines*, which are adaptable.

Cambridge illustrated dictionary for young scientists, Cambridge University Press 1985.

A handy source of reference for both adults and children.

Campbell, R J *Developing the primary school curriculum*, Holt Education, Eastbourne 1985.

DES *1987 Primary school staffing survey*, HMSO, London 1988.

Gilbert, C and Matthews, P *A guide to primary science policy*, Oliver & Boyd 1988.

Concerned with implementing Oliver & Boyd's *Look* scheme, but also generally useful.

Harlen, W *Guides to assessment in education science*, Macmillan Educational 1983.

Reviews ways of assessing science, test design and implementation.

Harlen, W *Primary science: taking the plunge*, Heinemann Educational 1985.

A starter book for the less confident.

Harlen, W *Teaching and learning primary science*, Harper & Row 1988.

Harlen, W and Jelley, S *Developing science in the primary classroom*, Oliver & Boyd 1989.

Clear and informative guidance on starting and developing scientific activities in primary schools; classroom organisation; excellent chapter on children's questions.

Hayes, M *Starting primary science*, Edward Arnold 1982.

Hodgson, B and Scanlon, E *Approaching primary science*, Harper & Row/The Open University 1985.

Part of the Open University unit on teaching primary science.

Johnsey, R *Problem solving in school science*, Macdonald Educational.
Activities aimed at challenging pupils' problem-solving abilities.

Jones, A *Things for children to make and do: science and technology activities*, Souvenir Press 1988.
Of special use to teachers of handicapped children.

Kingston encounters for young scientists.
Teachers' guide, air module and policy statement. From KPSU, Kingsdown Curriculum Centre, Ewell Road, Surbiton, Surrey KT6 6HL.

Lancashire looks at ..., *Science in the primary school* and *Science in the early years*.
Two resource books for teachers, with case studies and examples of activities. Also, a series of topic booklets (44 titles). Free catalogue from Lancashire Education Resources Unit, PO Box 61, County Hall, Fishergate, Preston PR1 8RJ.

Monaghan, H and Underwood, K *Early explorations*, Bedfordshire County Council 1988.
22 cards, teachers' book and folder for teachers of infants. Published by Bedfordshire Teaching Media Resources Service, Russell House, 14 Dunstable Street, Ampthill, Beds MK45 2JT.

NCC *Science in the National Curriculum*, HMSO, London 1989.

Peters, T and Austin, N *A passion for excellence*, Fontana, Glasgow 1985.

Raper, G and Stringer, J *Encouraging primary science*, Cassell 1987.
Practical advice on developing basic science teaching skills. Suggests ways of overcoming familiar difficulties.

Richards, C and Holford, D, eds. *The teaching of primary science: policy and practice*, Falmer Press.
A collection of articles concerned with primary science, mainly from *Education 3-13*.

Richards, R *An early start to science*, Macdonald Educational.

Ideas for primary and first school.

Science 5-16: a statement of policy, HMSO, London 1985.

Science for children with learning difficulties, Learning Through Science Project, Macdonald.
Selection from *Learning through science*; of specific use to teachers of slow learners.

Science starts here! series, Nelson 1988: *Making it move, Staying alive, Making changes* and *Sun, rain and myself.*
Useful starting points for junior teachers; full of practical ideas to be used with or without the Central TV series.

Showell, R *Primary science: a guide*, Ward Lock Educational 1983.
A resource book of topics. Ideas and activities with background information.

Science Schemes and Activity Packs

A first look! Oliver & Boyd.
For children up to about 8 years old. Consists of activity cards divided into five themes and a background guide.

Blueprints: 5–7 Science and *7–11 Science*, Stanley Thornes 1990.
Teachers' resource book with classroom-tested ideas for each attainment target at each level. Pupils' copymaster book provides 117 photocopiable pupil sheets of activities related to the attainment targets.

Chase, R *Argon and Xenon*, Blackwell 1990.
Science, design and technology activities based on a story-telling approach. Stories centre around the adventures of two light bulb characters, Argon and Xenon. Consists of teachers' resource book with ideas, activities and record sheets, picture book and activity cards; Key Stage 1.

Collins primary science, Collins, London 1990.
An activity and topic-based course which consists of ten pupils' books, grouped according to key stage. Each set has an accompanying assessment book and teachers' guide.

Fielding, D *Looking at science*, Blackwell.
A series of five combined science and nature study books for 8- to 12-year-olds.

Furnell, N *First science investigations*, Blackwell 1989.
20 experiments to develop scientific skills in the 9 to 12 age-group. Also contains teachers' notes, an assessment scheme and examples of children's work. Photocopiable in purchasing school.

Ginn science, Ginn.
Resource files at each level contain teacher's notes, activity support sheets, assessment material and help with planning, management and recording. Group discussion books at Levels 1 to 4, story and information books at all levels. Five of seven levels published (1990).

Jennings, T *The young scientist investigates*, Oxford University Press.
Science for 7- to 11-year-olds. Includes two course books, 20 topic books, teachers' manual and photocopiable teachers' book of practical work. OUP also publishes *Guides to the physical world* and *Living world*.

Kincaid, D and Coles, P *Footsteps into science*, Stanley Thornes.
Topic-based science series consisting of nine pupil books covering key primary topics, backed by a teachers' guide.

Learning through science, Macdonald Educational.
Guide, index and 12 themes of junior children's learning materials. Each theme consists of 24 assignment cards and a teachers' guide. Each pack contains two of each card.

Look! Oliver & Boyd.

Pack A is for younger children (7 to 9), Pack B for older children (9 to 11). Each pack contains a teachers' guide and 71 work cards.

New horizons: science 5–16, Cambridge University Press/West Sussex Science Education Development Unit 1990.

From the team that developed *Science horizons*. Primary component consists of teachers' resource packs and files, pupils' story and information books, games packs and flip books.

Newton, D P and Newton, L D *Footsteps into science*, Stanley Thornes.

A novel approach to junior science and technology. The scheme consists of 32 work cards and teachers' notes. Each card uses the achievements of a real scientist or technologist to stimulate children's interest in science and first-hand experience.

Rowlands, D and Holland, C *Problem solving in primary science and technology*, Stanley Thornes.

48 photocopiable worksheets providing open-ended activities in a ring binder, accompanied by a detailed teachers' guide. For the 5 to 11 age-range.

Scene setters, British Gas.

60 activity-based work cards with support material for the teacher in six units: *Air, Flight, Forces, Heat, Ourselves* and *Water*. Details from British Gas Education Service, PO Box 46, Hounslow TW4 6NF.

Science for primary schools, Ward Lock Educational.

A core science course for 7- to 11-year-olds developed around nine themes. Themes are organised into two pupils' books, a teachers' book and reproducible resource sheets at each of four levels.

Scienceworld: Science through infant topics, Longman.

Big books, four starter readers and three teachers' books which contain ideas to support the suggested activities.

Scienceworld. Junior teachers' and pupils' books 1 to 4, Longman.

Designed to follow *Science through infant topics*. Ten topics in each book. Teachers' books give background science and details of activities within topics.

The Science Collection, Mary Glasgow.

Information supplemented and supported by practical activities to help children cover the content of the National Curriculum. Five strands, each consisting of four 48-page books: *Ecology, Living Processes, Materials, Forces and Energy* and *Our World*.

Think and do: science for the early years, Blackwell.

Sets of 24 large, full colour cards. Children are encouraged to think about the problem posed by the card, to work out a strategy for achieving a solution, and to learn by doing. All activities are supported by teachers' notes on the reverse of each card.

ASE Publications

(See Useful Addresses list, page 73.)

A post of responsibility in science (revised 1981).

Be safe! Some aspects of safety in science and technology for primary schools.
An essential guide, with sections on 'Making things', 'Science out of doors', 'Electricity', 'Chemicals', 'Animals' and 'Heating and burning'. But check with your education authority guidelines as there may be variations.

Choosing published primary science materials for use in the classroom.
Popular schemes evaluated according to a set of criteria drawn up by the ASE's Primary Science Sub-Committee (updated 1990).

Implementing primary science education series.
The findings and recommendations of the IPSE evaluators of primary science implementation in three publications, *Report, The school in focus*, and *Building bridges*.

The headteacher and primary science (revised 1981).

The National Curriculum: making it work for the primary school.
A booklet written by teachers with special interest in science, mathematics and English for use by teachers in planning the cross-curricular implementation of the National Curriculum at Key Stage 1.

APU Reports for Teachers

Published by the DES and available from the ASE Publications Department (see Useful Addresses). The following is a selection:

No 4. Science assessment framework at age 11.
An account of the categories used for assessment in the APU surveys, with examples of questions and marking schemes.

No 6. Practical testing.
An account of the practical testing carried out at ages 11, 13 and 15 in the APU surveys.

No 8. Planning scientific investigations at age 11.
How the skills of planning investigations were assessed in the APU surveys; the implications for teaching.

Packs

Co-ordinating science in primary schools.
A video pack produced by IPSE and Newcastle LEA to form a training pack for primary science co-ordinators, comprising 11 workshop sessions and in-school work. Details from ASE (see Useful Addresses).

Magazines

Child education and *Junior education*.
Monthly journals for primary teachers which feature curriculum articles relating to science on a regular basis, plus pull-out theme sections and colour posters. For details, contact Subscription Dept, Scholastic Publications Ltd, Westfield Road, Southam, Leamington Spa, Warwickshire CV33 0JH. Also, *Infant projects* and *Junior projects*, bi-monthly resources which explore different cross-curricular projects incorporating science.

Primary science review.
Published three times a year. Contains articles, reviews and research findings relating to primary science. Available to those subscribing as individuals or as primary school members of the ASE.

Questions.
A magazine for primary teachers devoted entirely to science, design and technology, published three times a term. Features ideas for projects, guides to official documents, parents' page, reviews. Each issue includes a 16-page pull-out resource pack on a theme for the term, plus photocopiable activity sheets. Subscription enquiries to: Subscription Dept, *Questions* magazine, 6/7 Hockley Hill, Birmingham B18 5AA.

Useful Addresses

Association for Science Education (ASE), College Lane, Hatfield, Herts AL10 9AA.
The professional organisation for teachers of science in the UK. Primary teacher members receive: *Primary science review* five times a year; *Education in science* (the Association journal) five times a year; *ASE primary science*, a broadsheet published three times a year; access to science book selling service with 10 per cent discount; free indemnity insurance. The ASE Annual Meeting, which takes place in January, is the biggest science education event of the year.

British Association for the Advancement of Science, Fortress House, 23 Savile Row, London W1X 1AB.
Has a Youth Section (BAYS) which organises activities to complement and reinforce the curriculum, including award schemes for project work at primary level (Young Investigators); national and regional competitions; BAYS days and science and technology fairs. Also publishes a quarterly magazine, *Scope*, and produces support materials for teachers organising science clubs.

British Society for the History of Science, 31 High Street, Stanford in the Vale, Faringdon, Oxon SN7 8LH.
Has an education section for teachers.

British Telecommunications plc, Education Service, 81 Newgate Street, London EC1A 7AJ.

CLEAPSS (Consortium of LEAs for Provision of Science Services), School Science Service, Brunel University, Uxbridge UB8 3PH.
An information and advisory service, especially on equipment and safety – check to see if your local authority is a member.

Council for Environmental Education, University of Reading, London Road, Reading RG1 5AQ.
A national centre for environmental education. Provides an information and advisory service and publishes resource sheets for teachers. A Youth Unit works outside the formal education sector with organisations concerned with the environmental education opportunities of young people.

Earth Science Teachers Association, c/o Sheila Rogers (Membership Secretary), 4 Middledyke Lane, Cottingham, North Humberside HU16 4NH.
Encourages and supports the teaching of earth science. Members receive *Teaching earth science* four times a year.

Equal Opportunities Commission, Overseas House, Quay Street, Manchester M3 3HN.
For information and educational materials on women scientists.

Friends of the Earth, 26–28 Underwood Street, London N1 7JQ.

General Dental Council, 37 Wimpole Street, London W1M 8DQ.

Health Education Council, 78 New Oxford Street, London WC1A 1AH.

Meteorological Office, London Road, Bracknell, Berks RG12 2SZ.

Centre for Alternative Technology, Machynlleth, Powys SY20 9AZ.

National Curriculum Council, Albion Wharf, 25 Skeldergate, York YO1 2XL.
Contact for information on all aspects of the National Curriculum.

Nature Conservancy Council, Northminster House, Northminster Road, Peterborough PE1 1UA.
Provides advice and support for education on conservation and financial support for wildlife areas. *Points of view* leaflets introduce and explain debates about controversial issues.

Royal Society for Nature Conservation, The Green, Witham Park, Waterside South, Lincoln LN5 7JR.
The majority of RSNC's educational work is achieved through WATCH – the wildlife and environment club for children. Services are offered to schools through project-based materials.

Royal Society for the Prevention of Cruelty to Animals, Causeway, Horsham, West Sussex RH12 1HG.

The RSPCA produces a range of relevant materials, for example, manuals on the care of animals kept in schools, as well as videos and slides.

Royal Society for the Protection of Birds, Education Department, The Lodge, Sandy, Beds SG19 2DL.

Produces a whole range of resources for primary schools, including *Bird studies for primary science in the National Curriculum: a guide to practical studies*.

Science and Technology Regional Organisation (SATRO), 76 Portland Place, London W1N 4AA.

SATRO operates in 42 locations throughout the UK and aims to enhance young people's understanding of science, engineering, industry and technology through closer working relationships between schools and the outside world.

Scottish Schools Science Equipment Research Centre, 24 Bernard Terrace, Edinburgh EH8 9NX.

Scottish subscribers only.

STEEL, Woodlands Centre, Southport Road, Chorley, Lancashire PR7 1QR.

STEEL (Lancashire's SATRO) produces a series of primary science publications.

The Tree Council, 35 Belgrave Square, London SW1X 8QN.

The Wildfowl and Wetland Trust, Education Officer, Slimbridge, Gloucestershire GL2 7BT.

World Wide Fund for Nature, Panda House, Weyside Park, Cattershall Lane, Godalming, Surrey GU7 1XR.

Zoological Society of London, Regents Park, London NW1 4RY.

Places to Visit

British Museum (Natural History), Cromwell Road, London SW7 5BD. Tel: 071 589 6323.

Interactive or hands-on science centres feature specially constructed exhibits that encourage visitors to investigate natural phenomena and experiment with technology. Guides or enablers are usually on hand to help if required. The following centres all have education contacts on the staff and a booking system for school parties.

Discovery Dome, travelling interactive centre, c/o The Nuffield Foundation, 28 Bedford Square, London. Tel: 071 631 0566.

Green's Mill and Centre, Belvoir Hill, Sneinton, Notts. Tel: 0602 503635.

Jodrell Bank Science Centre, Macclesfield, Cheshire. Tel: 0477 71339.

Launch Pad, Science Museum, Exhibition Road, London. Tel: 071 938 8000.

Light-on Science! Birmingham Museum of Science and Industry, Newhall Street, Birmingham. Tel: 021 236 1022.

The Micrarium, The Crescent, Buxton, Derbyshire. Tel: 0298 78662.

Techniquest, Britannia House, Britannia Road, Cardiff. Tel: 0222 460211.

Technology Testbed, Large Objects Collection, Princes Dock, Merseyside County Museum, Liverpool. Tel: 051 236 0642.

The Exploratory, Bristol Old Station, Temple Meads, Bristol BS1 6QU. Tel: 0272 252008.

Xperiment! Greater Manchester Museum of Science and Industry, Manchester. Tel: 061 832 2244.

The Science Museum, Exhibition Road, London SW7 2DD.

Open all year round except 25 Dec, 1 Jan and 1 May. For information on booking and facilities for schools, telephone 071 938 8222.

Other Resources

Early SATIS (Science and Technology in Society).

A resource to place science and technology in a social context. Uses a range of tried and tested active learning approaches to involve upper primary and lower secondary pupils in National Curriculum related issues. First materials available from the ASE, January 1992.

NERIS (National Educational Resources Information Service).

A curriculum database service for schools which provides information about learning materials, their availability and, where appropriate, the materials themselves for downloading and printing in school. Information is also available about places to visit and speakers to contact. For details, contact NERIS, c/o Maryland College, Leighton Street, Woburn MK17 9JD.

MIST (Modular Investigations into Science and Technology).

A videodisc resource for the staffroom and classroom. For details, contact MIST, Evergreen Communications, 2nd floor, 5 Dean Street, London W1V 5RN.

Pictorial Charts Educational Trust, 27 Kirchen Road, London W13 OUD.

Produces wallcharts and posters to link with science and technology, biology and health and the human body. Most charts are accompanied by a booklet of copyright-free background notes.

Softlab: Shell software guides to laboratory work.

A store of equipment related to various topics which may be put together on screen. Intended to extend the number and scale of scientific ideas which may be explored by pupils. Titles include: *Electricity, Floating and sinking* and *Hot and cold*. Each guide consists of teachers' notes and disk (BBC Model B and Master). Order direct from Hodder and Stoughton Educational, Mill Road, Dunton Green, Sevenoaks, Kent TN13 2YD.

Viewtech Audio Visual Media, 161 Winchester Road, Brislington, Bristol BS4 3NJ.

Has a whole range of films, videos, filmstrips and slide sets available for purchase or hire on primary science related themes.

Suppliers

A & L Scientific and Optical Equipment, 190 West Drive, Clevelys, Blackpool, Lancashire FY5 2EJ.
Low cost equipment, especially magnifiers and binocular microscopes.

E J Arnold and Son Ltd, Dewsbury Road, Leeds LS11 5TD.
Primary school equipment.

Commotion, 241 Green Street, Enfield EN3 7TD.
Especially technology.

GeoPacks, c/o MJP, PO Box 23, St Just, Cornwall TR19 7JS.

Griffin and George Ltd, Bishop Meadow Road, Loughborough, Leicestershire LE11 ORG.
Specialist supplier.

Philip Harris, Lynn Lane, Shenstone, Lichfield, Staffordshire WS14 OEE.
Specialist supplier.

Heron Educational Ltd, Unit 12, Kenilworth Works, Denby Street, Sheffield.
Technology equipment.

Hestair Hope, St Philip's Drive, Royton, Oldham OL2 6AG.
Primary school equipment.

Hogg Laboratory Supplies, Sloane Street, Birmingham B1 3BW.
Specialist supplier.

NESTEC Ltd, Unit 24D, North Tyne Industrial Estate, Whitley Road, Longbenton, Newcastle NE12 9SZ.

Osmiroid International Ltd, Fareham Road, Gosport, Hampshire PO13 OAL.
Magnifiers, nets, Teko construction material.

Nottingham Educational Supplies, 17 Ludlow Hill Road, West Bridgford, Nottingham.
General primary, but especially construction kits.

C E Offord, Hurst Green, Etchington, East Sussex.

Sci Tech Educational, PO Box 190, New Dover Road, Canterbury, Kent CT1 3BH.

Surplus Buying Agency, Birley School, Fox Lane, Sheffield S12 4WU.
Low cost small parts and consumables.

Technology Teaching Systems, Penmore House, Hasland, Chesterfield, Derbyshire.
Especially technology.